HAMMER, SICKLE, AND SOIL

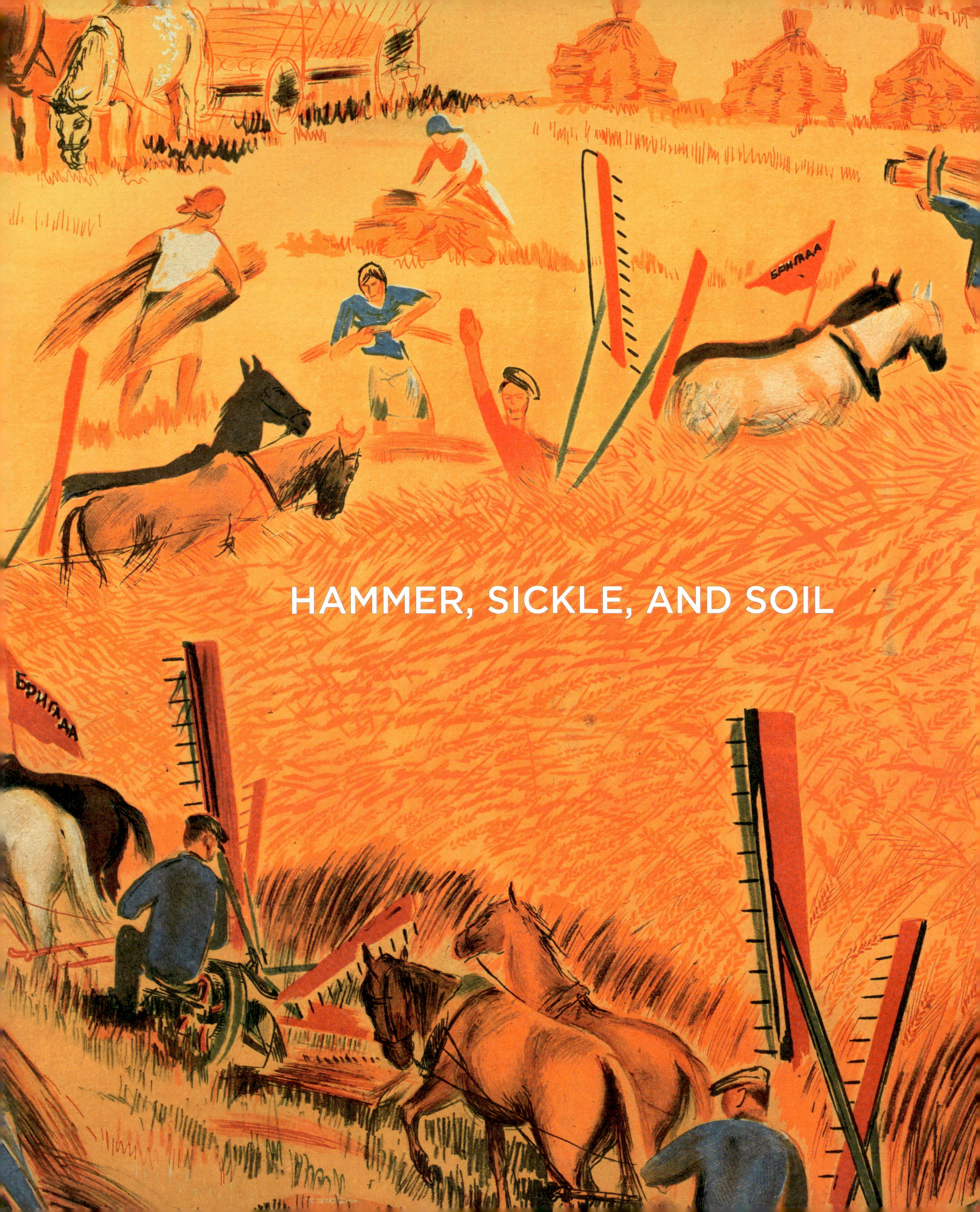

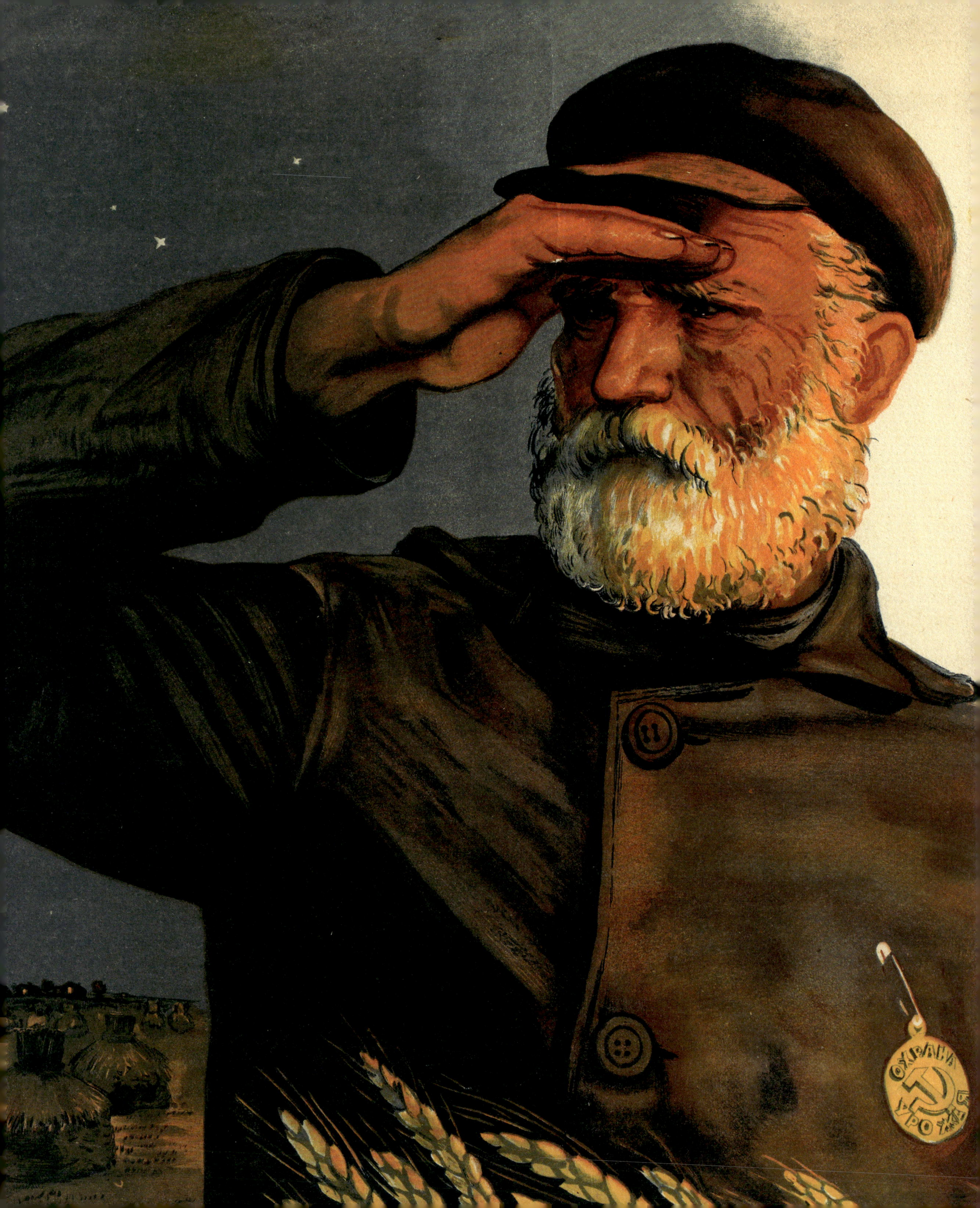
ОХРАНА
УРОЖАЯ

HAMMER, SICKLE, AND SOIL

The Soviet Drive to Collectivize Agriculture

Jonathan Daly

HOOVER INSTITUTION PRESS
Stanford University | Stanford, California

With its eminent scholars and world-renowned library and archives, the Hoover Institution seeks to improve the human condition by advancing ideas that promote economic opportunity and prosperity, while securing and safeguarding peace for America and all mankind. The views expressed in its publications are entirely those of the authors and do not necessarily reflect the views of the staff, officers, or Board of Overseers of the Hoover Institution.

www.hoover.org

Hoover Institution Press Publication No. 680
Hoover Institution at Leland Stanford Junior University,
Stanford, California, 94305-6003

First printing 2017
24 23 22 21 20 19 18 17 9 8 7 6 5 4 3 2 1

Designed by Jennifer Navarrette, with additional design and production assistance by Wilsted & Taylor Publishing Services, Oakland, California.
Printed and bound by Friesens Corporation in Canada.

The paper used in this publication meets the minimum requirements of the American National Standard for Information Sciences—Permanence of Paper for Printed Library Materials, ANSI/NISO Z39.48-1992. ∞

Cataloging-in-Publication Data is available from the Library of Congress.
ISBN 978-0-8179-2064-7 (cloth : alk. paper)
ISBN 978-0-8179-2066-1 (EPUB)
ISBN 978-0-8179-2067-8 (Mobipocket)
ISBN 978-0-8179-2068-5 (PDF)

PAGE i Detail of "Let us expand socialist competition and shock work," *see page 77.*
PAGE ii Detail of "Vigilantly protect the socialist harvest," *see page 74.*

To Christine, Anna, Zoe, and Jeremiah,

hoping for a better world

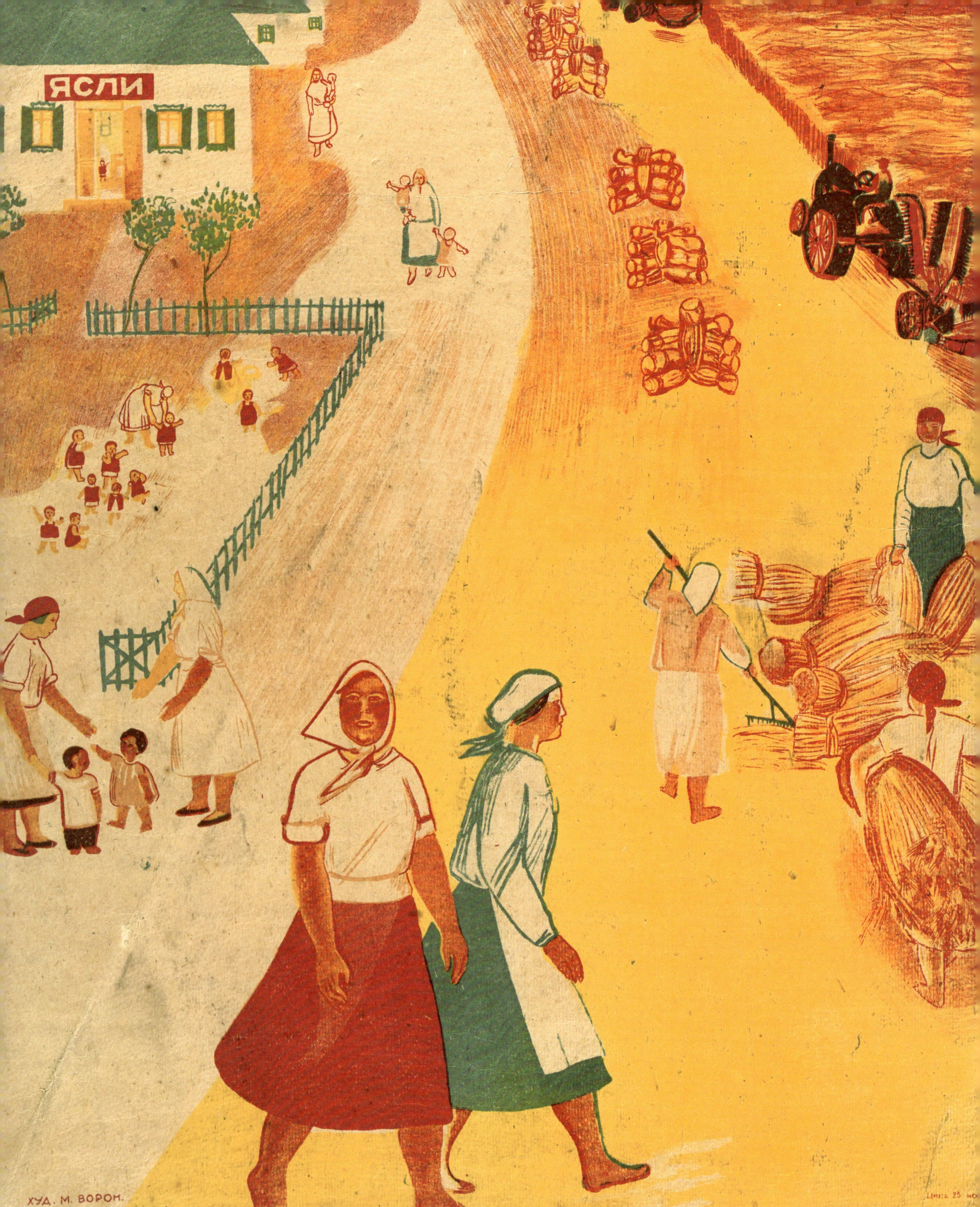

ХУД. М. ВОРОН.

ОГИЗ–ИЗОГИЗ

Цена 25 ко

CONTENTS

◀ Detail of "Collective farm women," *see page 98.*

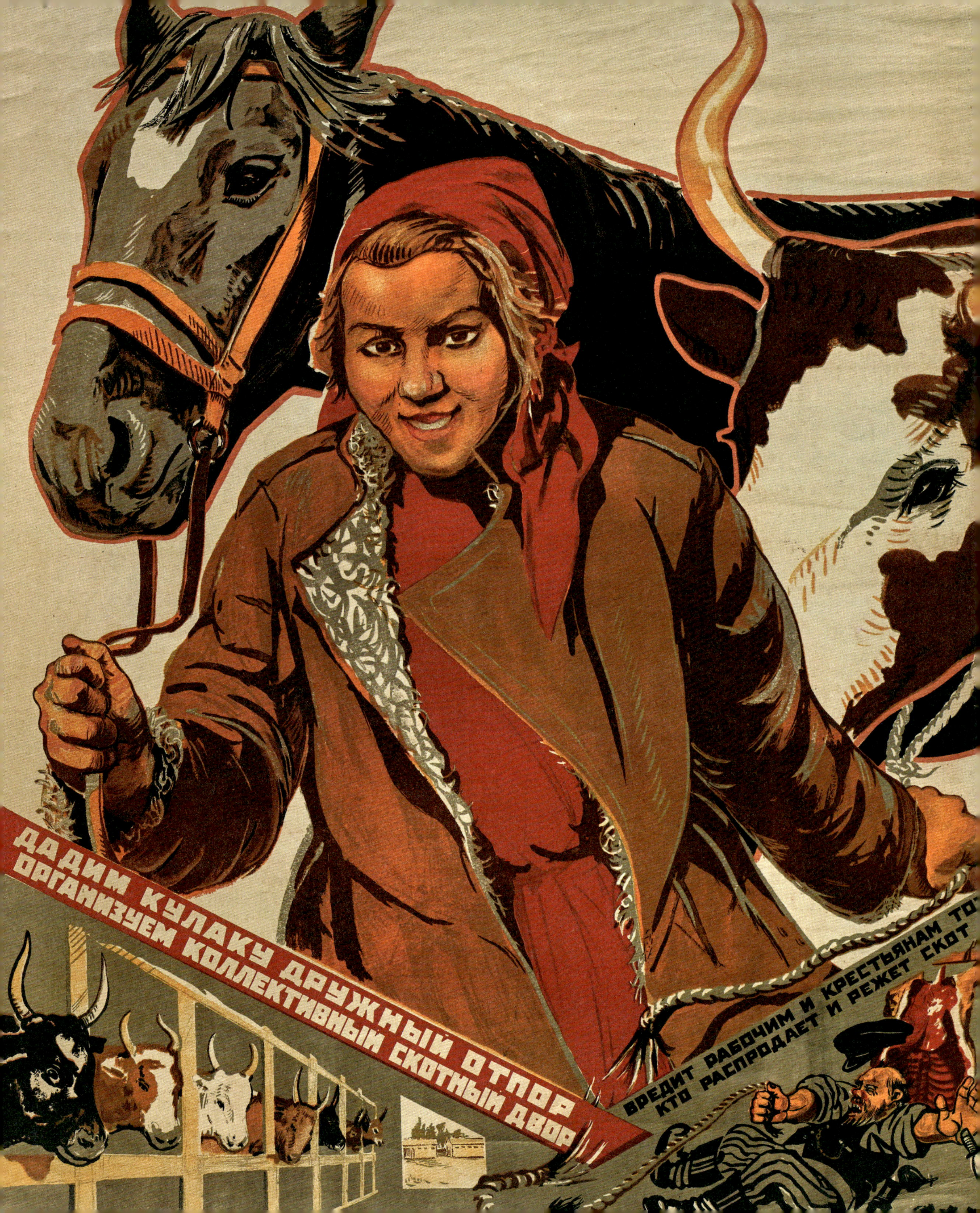
ДАДИМ КУЛАКУ ДРУЖНЫЙ ОТПОР
ОРГАНИЗУЕМ КОЛЛЕКТИВНЫЙ СКОТНЫЙ ДВОР
ВРЕДИТ РАБОЧИМ И КРЕСТЬЯНАМ ТОТ
КТО РАСПРОДАЕТ И РЕЖЕТ СКОТ

ILLUSTRATIONS

◂ Detail of "Join the collective farm," *see page 43*.

▶ Detail of “Let us create indivisible funds,” *see page 31*.

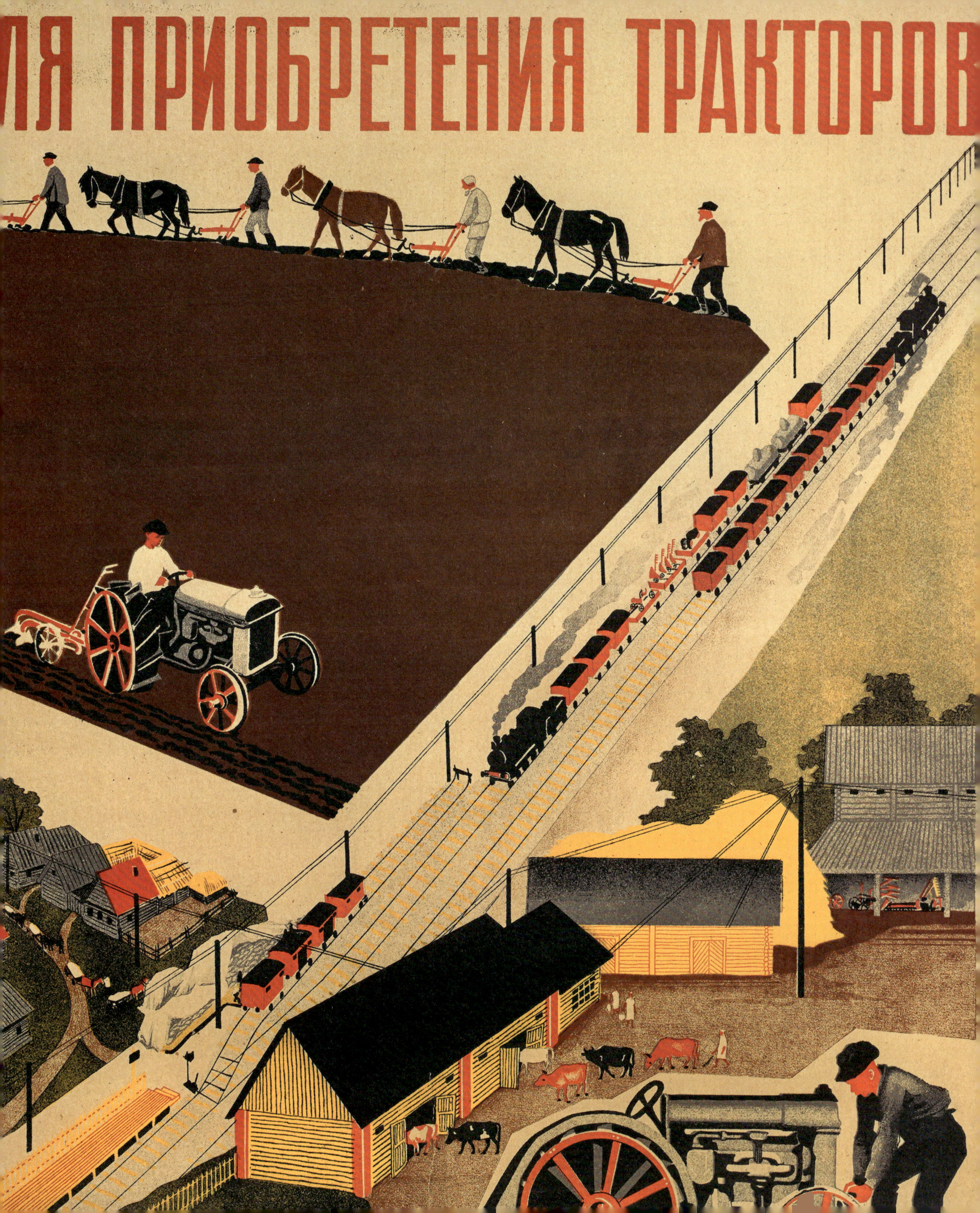
ЛЯ ПРИОБРЕТЕНИЯ ТРАКТОРОВ

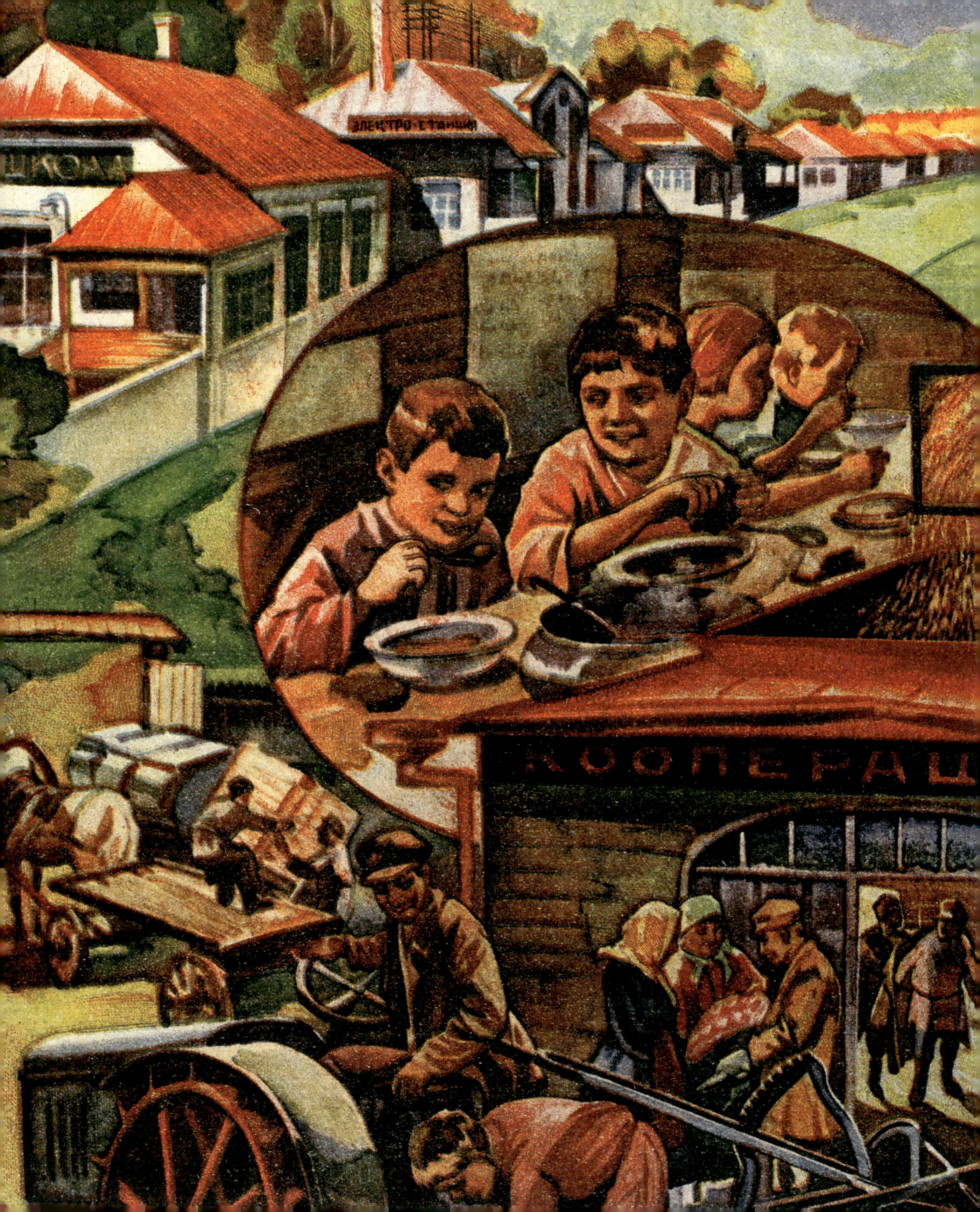
ШКОЛА
ЭЛЕКТРО-СТАНЦИЯ
КООПЕРАЦ

FOREWORD

It is a tremendous pleasure for the Hoover Institution to support the publication of Professor Jonathan Daly's illustrated history of the collectivization of agriculture in Soviet Russia. Not only does *Hammer, Sickle, and Soil* narrate what arguably was one of the most significant—and devastating—policies put forward during Stalin's rule in Russia, but it is thoughtfully designed to incorporate more than sixty images from the Hoover Institution Library & Archives' world-renowned collection of Soviet posters—posters that are critical to illuminating the ideology of one of the world's cruelest authoritarian regimes as it enforced large-scale agricultural reform upon its citizens. The wide gap between the poster images of young worker-heroes behind tractor wheels or bringing in idyllic harvests and the reality of the brutal experience of Russian peasants under Stalin's collectivization tells a crucial story in Soviet Russia's history. Professor Daly's work represents a substantial step toward our scholarly understanding of Soviet collectivization and the role it played in the demise of the Soviet socialist experiment.

Professor Daly's work is the product not only of research in Hoover's vast Russian holdings—generally considered to be the most significant outside of Russia itself—but also of his participation in Hoover's annual Workshop on Totalitarian Regimes. For more than a decade, this workshop has gathered students and faculty from across the globe to study, discuss, and present findings on the history, structure, and legacies of authoritarian states past and present. Although in its beginning the workshop focused on Soviet Russia, it has recently been expanded to include projects on China under Mao, Iraq under Saddam Hussein, and other authoritarian states. The study of Soviet Russia, however, remains crucial to research in the Hoover reading room and to the workshop; the Russian collections at Hoover have proved critical to the works of such well-regarded historians as Robert Conquest, Stephen Kotkin, Norman Naimark, and Robert Service. Hoover is pleased to support the work of both established luminaries in the field and young scholars as they conduct vigorous, meaningful research that furthers our understanding of the history of human action.

In terms of the historical record, it seems most appropriate for the Hoover Library & Archives to support a project on the history of famine in Russia. Our founder and namesake, Herbert Hoover, building on the masterful relief organizing he executed for Belgium in World War I, led the American Relief Administration (ARA) in a massive effort to feed the Russian peasants during the horrific famine of 1921–23. As Professor Daly notes in his introduction to *Hammer, Sickle, and Soil*, without the assistance of the ARA, millions more than the

◀ Detail of "Down with moonshine!," *see page 116*.

five million Russians who are estimated to have died would have perished from starvation and disease during the famine. In all, the ARA distributed more than $60 million worth of food and medical supplies from Europe and the United States to the devastated Russian countryside. The atrocities and starvation suffered by civilians in the World War I and postwar era inspired Hoover in 1919 to build a repository of records, photographs, documents, maps, and ephemera where scholars from around the world could gather to study war, revolution, and peace, yielding research that would lead to a more peaceful world. The Hoover Library & Archives continue to honor the founder's mission by collecting, preserving, describing, and making available to the public the world's most important materials on social, political, and economic change in the twentieth and twenty-first centuries.

Hoover's collection of more than three thousand Soviet posters, featured in *Hammer, Sickle, and Soil*, is one of the best-known of such poster collections in the world and is frequently consulted as a source for the study of art, political propaganda, graphic design, typography, and the history of Russian socialist ideology. As art, the posters represent the Soviet Central Committee's desire to do away with traditional caricaturist Russian images and create a new, bold, proletarian art—one that would often be rendered by well-known avant-garde artists such as Vladimir Mayakovsky, Demyan Bedny, Alexander Rodchenko, and El Lissitsky. Embracing bold experiments in text, font, color, and geometry, the Central Committee printed millions of posters that were displayed in all public areas: on trains, in streets, on fences, in factories, offices, cafeterias, and schools. The messages conveyed by the posters addressed literacy and hygiene campaigns, economic development, political indoctrination, social change, and political education. The posters designed to promote agricultural collectivization, however, are among the rarest and most interesting subsets of the canon of Soviet posters.

As ephemera—but particularly as ephemera posted in impoverished, starving regions—the posters often disappeared into dustbins, fireplaces, or rubble, making them rare artifacts. The collectivization posters also represent one of Stalinist Russia's most desperate attempts to gain control of public discourse in regions largely hostile to urban bureaucrats who arrived in the countryside to carry out national policy and law. Collectivization gave rise to new symbols and imagery aimed specifically at the peasant, who was often perceived by Soviet leaders as superstitious, uneducated, religious, and hostile to the revolution. Before Stalin's Five-Year Plan, the visual syntax of Soviet posters included the sickle, hammer, and red star. After collectivization was launched, these images were joined by the tractor, the slim female peasant, the "kulak" (rich, land-owning peasant) as a fat, surly miser, demonized clerics, and the image of Stalin overlooking rich harvests of grain. With the advent of agrarian reform came a whole new language of visual iconography—one often simplified for the low level of literacy in rural areas. Taken in sum, the collectivization posters provide a rich history of communist aspirations for progress, industrialization, and economic success in the 1920s and 1930s; they also provide a telling, remarkable foil for the actual economic devastation wrought by the Five-Year Plan. As Professor Daly reminds us, the Soviet agricultural collectivization effort and the anti-"kulak" campaign laid the groundwork for the gulag system that would haunt Soviet Russia for the remainder of the century.

When researching in Hoover's Soviet poster collection, one notices that many of the posters contain the warning that "Anybody who tears down or covers this poster is committing a counter-revolutionary act." We can only hope that these posters have been rescued and preserved for those very counterrevolutionary reasons: so we and future generations can study authoritarian regimes, learn from history, and build pathways toward a more peaceful world. Professor Daly's *Hammer, Sickle, and Soil* provides a significant step forward in those efforts; the Hoover Library & Archives are proud to support the endeavor.

—*Eric Wakin, Robert H. Malott Director of Hoover Institution Library & Archives, Hoover Institution deputy director, and research fellow*

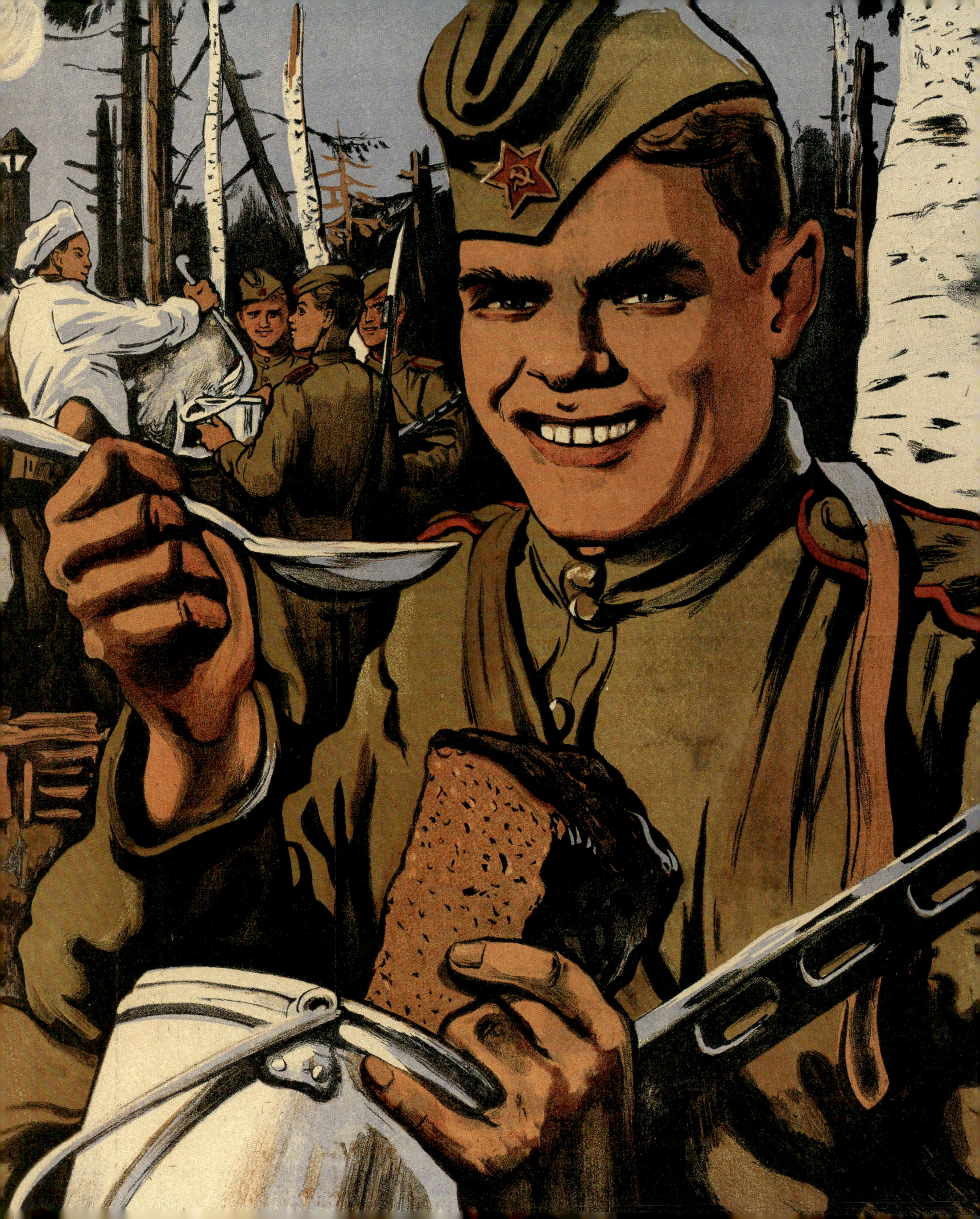

PREFACE

Their houses are very ill built, the walls bevel, without one right angle in any apartment; and this defect arises from the contempt they bear to practical geometry, which they despise as vulgar and mechanic.

—Jonathan Swift, *Gulliver's Travels*

This book tells the harrowing story of Stalin's transformation of some 26 million family farms throughout the USSR into approximately 250,000 collective farms in 1929–33. In this process, roughly 12 million people left the countryside for urban areas, peasants slaughtered over 40 million horses and cattle (or 40 percent of the total numbers) to avoid confiscation, between 2 and 2.5 million peasants were relocated within their districts, another 2.1 million were herded into boxcars and exiled to remote localities, and at least 5 million perished in the famine of 1932–33. This was, at that time, history's biggest experiment in social engineering and the first example of the complete conquest of the bulk of a population by its own rulers. It was also above all intended to bring to Russia Marx's promised bright future of socialism.

While attending the Workshop on Totalitarian Regimes at the Hoover Institution in July 2015 and conducting research in its superb collections of Soviet documentation, I discovered an extraordinary cache of propaganda posters relating to the Soviet collectivization drive. It was clear that these posters could marvelously illustrate the drama and trauma of the revolution in Soviet agriculture under Stalin. All the illustrations in this book come from the Hoover Archives.

Of course, the reality of collectivization often diverged radically from the optimistic and even joyful propaganda posters, which were for the most part created in the style and according to the philosophical tenets of socialist realism. This official artistic movement aimed to represent the achievements of Soviet socialism in the light of their future fulfillment—not life as it was then experienced, but as it should be, were everything to go according to plan, or as it would be in the "bright socialist future." Such propaganda posters were a ubiquitous everyday feature of Soviet popular culture, both adornments of public spaces and didactic tools for raising consciousness and imparting knowledge about the world to Soviet citizens. Thus, for example, a major theme of the collectivization posters is women eagerly rushing to join the collective farm, though in reality women constituted the bulk of the peasants protesting forced collectivization (the peasants knew that women were less likely than men to face serious punishment). Women were not alone in resisting the collectivization drive. Peasant unrest reached massive proportions: in 1930 alone, according to official statistics, 1,198 activists and officials promoting collectivization were killed, and roughly 5,000 more were assaulted, while nearly 14,000 "disturbances" involving 2.5 million participants were officially counted. The accompanying text will fill in these and other blanks to provide a full and accurate account.

Thanks to the huge outpouring of scholarly studies and primary source collections of the past two decades, nearly every contour and aspect of the story is

◀ Detail of "To advanced collective farms—thank you from the front," *see page 127.*

now known.[1] Yet there is in English neither an up-to-date scholarly survey of these developments nor a readily accessible but accurate narrative for the general reader. While both are needed, it is my intention with this volume to fill the second gap.

Paul Gregory and Mark Harrison, the director and codirector of the Workshop on Totalitarian Regimes, kindly encouraged me to pursue this project, for which I am grateful. My thanks go also to Eric Wakin, Deputy Director, and Robert H. Malott, Director of Library and Archives at Hoover, for approving and supporting the publication of this book, and the wonderfully efficient and helpful Hoover Press production team. In particular, it was a real pleasure to work with managing editor Barbara Arellano, production manager Marshall Blanchard, and copyeditor Elizabeth Berg. An LAS Award for Faculty Research in the Humanities from the University of Illinois at Chicago generously supported my research. I am also grateful to Wasif Ahmad for enthusiastic research assistance.

I dedicate this book to my children: Christine, Anna, Zoe, and Jeremiah—my pride, my joy, my hope. I am grateful for what you bring to my life.

◆ ◆ ◆

Russian is a Slavic language that uses the Cyrillic alphabet. To transliterate from Cyrillic into Western languages is tricky and requires certain conventions. I adhere to them with some exceptions. To write Trotskii, as required by the Library of Congress system of transliteration, instead of Trotsky, for example, would be confusing and therefore unnecessary.

HAMMER, SICKLE, AND SOIL

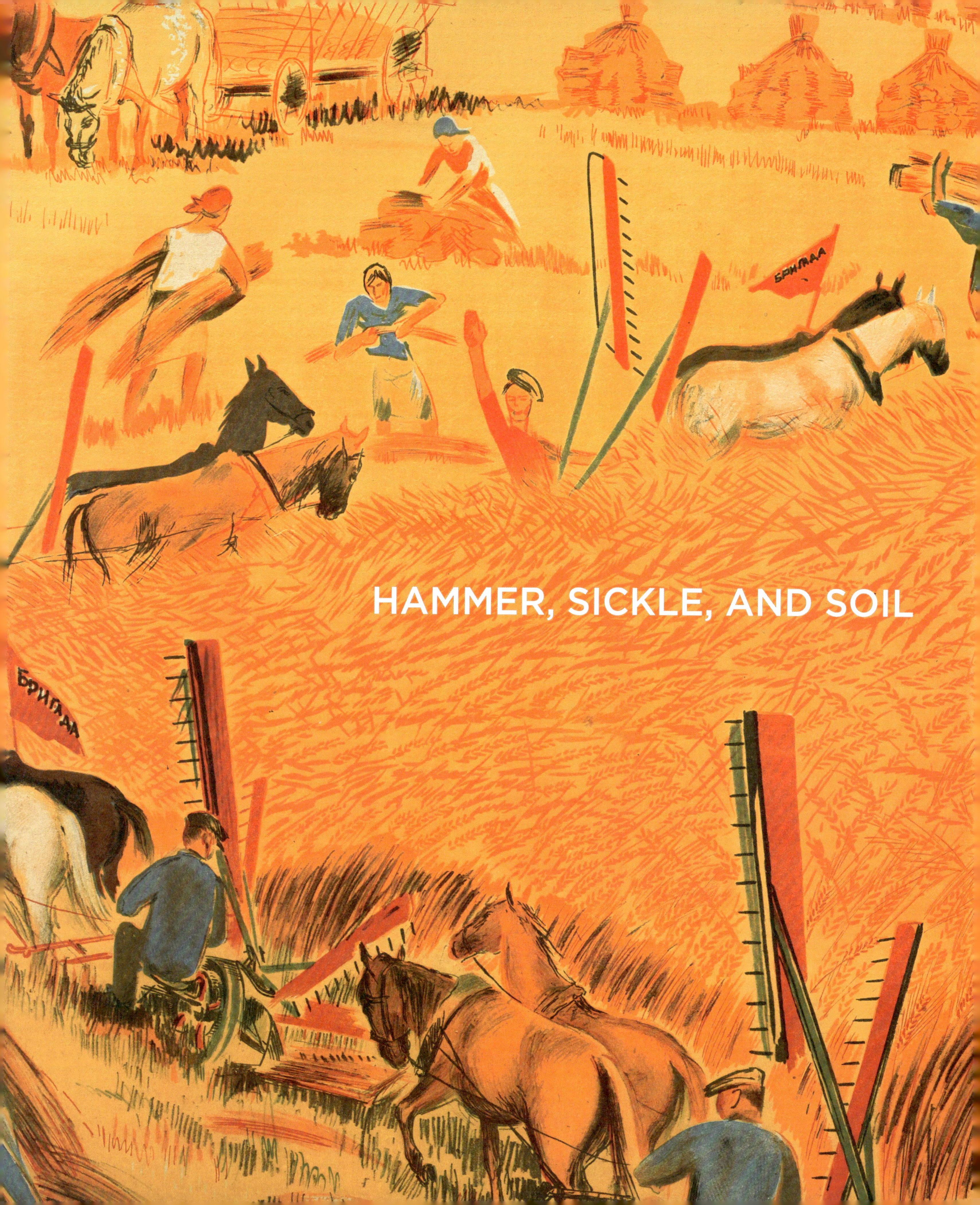

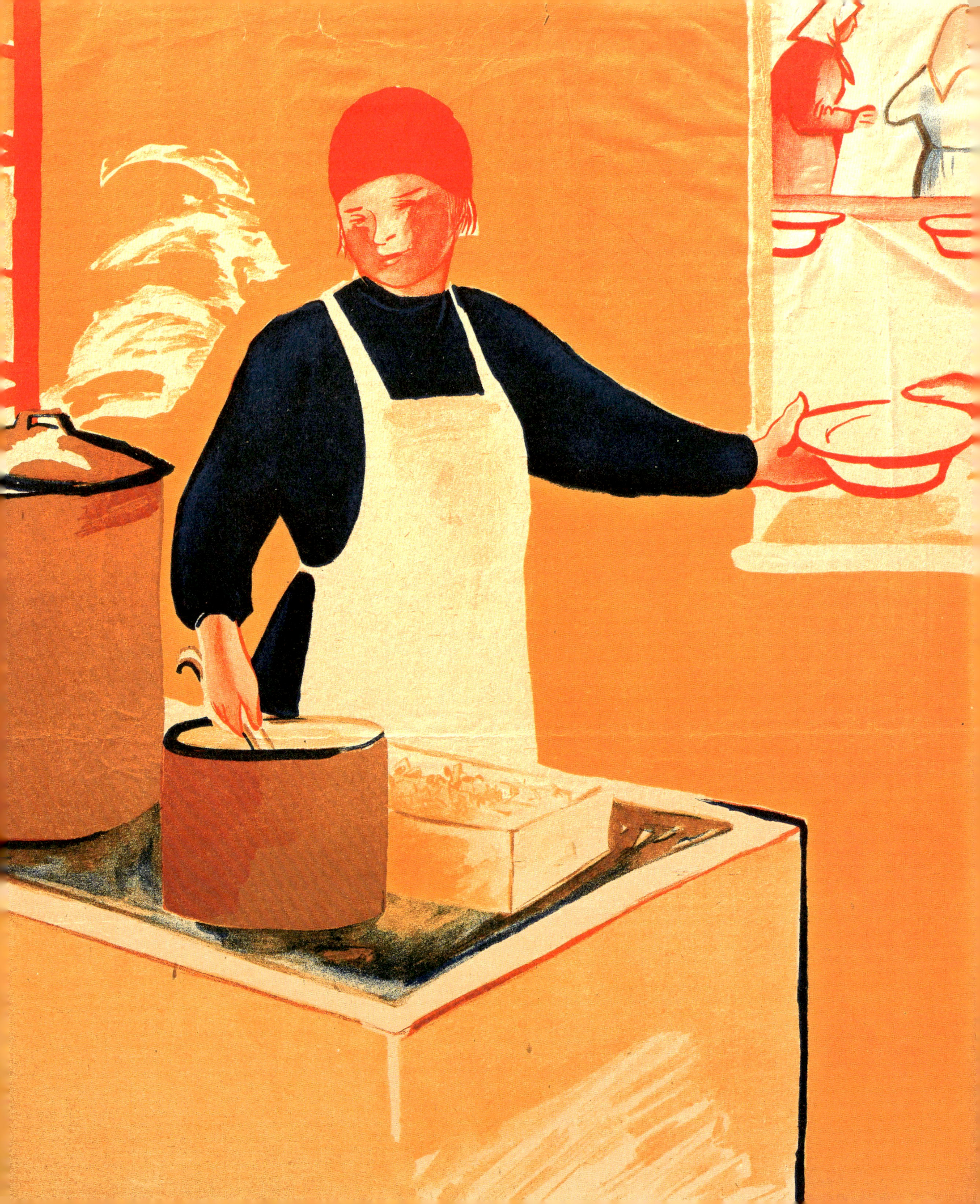

INTRODUCTION

Russia entered the modern age during the reign of Peter the Great (r. 1682–1725). Among his first acts as sole ruler, after his half-brother and coruler, Ivan V, died in 1696, was to spend eighteen months in Europe, learning everything he could from a continent in the dramatic throes of a transformation more radical than any region of the world had ever undergone. The printing revolution, the Renaissance and Reformation, the age of exploration and empire building, the scientific revolution, a commercial revolution involving transoceanic trade and stock exchanges, the Dutch War of Independence, the English Glorious Revolution—all had unleashed the creative skills and abilities of millions of Europeans, raised standards of living, begun unlocking the secrets of nature, empowered individuals in matters of faith and politics, and soon would launch the Industrial Revolution and the making of the modern world. Peter loved Europe's dynamism and the inventiveness of its people and resolved to remake his country in its image.

He studied a dozen trades—shipbuilding, dentistry, mapmaking, navigation, military science, engineering, and surgery, among others—and brought back with him specialists in these and other fields. He also sent hundreds of young Russians to study in Europe. Over the next quarter century, Peter built up a European-style bureaucracy and diplomatic corps, modernized the army, created a navy, instituted state control over the church, commissioned detailed maps of Russia, founded an entirely new city as his capital (Saint Petersburg), introduced European artistic and architectural methods and styles, issued decrees and political treatises explaining in rational terms his goals and purposes, dramatically expanded printing, simplified the Cyrillic alphabet, sponsored the publication of handbooks of all kinds, expanded education, inculcated new ways of speaking and thinking, promoted European manners of dress and grooming, established an Academy of Science and an Academy of Arts—one could go on and on.

The point is that in bringing about these staggering changes, Peter was setting up Europe as a model for Russia, thus making it the world's first developing country. Russia became more dynamic, its elites better educated, its military more powerful. Within a century and a half, Russia's greatest composers, writers, and scientists would rival those of any country in the world. At the same time, however, it perpetually fell short of its geopolitical rivals in military power, commercial prowess, and technological and industrial achievement. A thin layer of Europeanized elites grew apart from the bulk of the population, who remained tied to the land and wedded to premodern values and culture. To finance his grandiose projects, Peter promoted industrial development but also squeezed more taxes from the peasantry, the majority of whom were serfs. The state reigned supreme, jealously guarding its power,

◂ Detail of "In the collective farm, build a collective lifestyle," *see page 14*.

refusing to share authority even with the educated elites. Yet many political ideas emanating from Europe in the decades following Peter's death emphasized power sharing, checks and balances, popular participation, and social justice. Russia's rulers tried to shut out such ideas, while the intelligentsia—critically minded intellectuals emerging in the first half of the nineteenth century—zealously embraced them. Lacking institutions or rights enabling them to influence government policy, radical intellectuals treated European doctrines of social and political change as infallible guides to political understanding and action.

Russia's rulers also tried to keep up with the latest European developments. Catherine the Great (r. 1762–96) corresponded with Voltaire and borrowed ideas from the French constitutionalist Montesquieu and the English jurist Blackstone. Her grandson, Alexander I (r. 1801–25), seriously considered granting a constitution to Russia and later marched into Paris with a grand coalition against Napoleon. His grandson, Alexander II (r. 1855–81), freed over 23 million serfs and created an independent judiciary—entirely by autocratic fiat. He was, however, assassinated by revolutionary activists for whom his reforms had not gone far enough. His son, Alexander III (r. 1881–94), crushed the radical opposition and launched Russia on a career of industrialization. Levying heavy taxes and exporting grain to buy machinery and build up gold reserves helped bring about a horrific famine in 1891–92, which killed roughly four hundred thousand people (a figure that would have been far higher without the government's extensive relief efforts). The fact that Russia suffered such a calamity in an age when famines typically struck only colonial states like Ireland and India inspired public distrust and criticism. During the rest of Alexander's reign and that of his son, Nicholas II (r. 1894–1917), intellectual discontent festered and swelled.

Marxism enjoyed particular favor among Russia's intelligentsia because it purported to scientifically explain social and economic development in all countries and seemed to promise a path to the highest stage of human flourishing in the face of an apparently unshakable autocratic government and without the gradual and spontaneous emergence of political, social, and economic institutions witnessed in Europe. Neither the great mass of ignorant peasants nor authoritarian administrative officials and police, according to Russian interpretations of Marxism, could stop the march of history toward capitalism, through socialism, and ultimately to communism. In this drama, industrial workers, as participants in the most advanced sectors of the economy, would play the leading role.

The two main currents of Russian Marxism divided on the question of leadership. The Russian Social Democratic Workers Party split in 1903 into Bolsheviks, who insisted on centralized party control, and Mensheviks, who were more willing to trust historical forces and the class consciousness of the workers, meaning their alleged instinctive tendency toward socialist revolution. The leaders of both factions were westernizers, staunch believers in a universal path of progress, who rejected the idea of some special Russian path. They also, as consistent Marxists, mistrusted the peasantry as "petty bourgeois." The chief Bolshevik leader, Vladimir Ulyanov, known as Lenin (1870–1924), was by all accounts fanatical in his Marxist faith, resolutely authoritarian, ruthless, single-mindedly committed to revolution, deeply self-righteous, and absolutely confident in his political judgment.

Beginning in the 1890s, thanks in part to government encouragement of railroad construction and foreign investment,

Russia boasted one of the fastest-growing economies in the world. On the eve of World War I, its gross industrial output was fifth in the world and fourth in Europe. At the same time, its per capita national income was by far the smallest of the major European economies: in 1913, less than a third that of the United Kingdom and roughly half that of Austria-Hungary. Its population growth was extremely rapid, doubling over the course of the nineteenth century, increasing land hunger among the peasantry. The share of the urban population grew, too, from 6 percent in 1861 to 18 percent in 1913—a tremendous increase, though still a very low total relative to the major European countries (by comparison, 39.5 percent of the French population lived in towns in 1896). Russia was experiencing extremely fast development while remaining relatively backward. Such rapid change fueled mounting social tension.

Popular discontent broke out in a series of disorders beginning with major industrial strikes in 1896, student protests in 1899–1901, peasant uprisings in 1902, and anti-Jewish pogroms in 1903. Russia had been pursuing an expansionist foreign policy in the Far East, provoking the Japanese to attack the Russian Eastern Fleet at Port Arthur, which Russia had leased from China. In the course of the war, which went badly for Russia, mass unrest gripped the country, reaching a crescendo in October 1905 with a general strike that nearly shut down the economy. Nicholas II responded with the October Manifesto, promising civil rights and a representative institution. He partly went back on his promises but still granted a limited right to assembly, diminished censorship controls, and a bicameral legislature without whose approval few laws could be adopted.

Over the next several years the economy continued to expand, Russian culture flourished and took the world by storm, the peasantry dramatically expanded its landholdings at the expense of landowners, agricultural output increased by one-third (in 1900–1913), exports of grain rose by 50 percent (in the same years), literacy increased rapidly (73 percent of military recruits could read by 1913), and nearly a quarter of peasant households had taken advantage of a 1906 law permitting them to leave their rural communes and set up individual farms. Major revolutionary parties, including the Bolsheviks, had collapsed, from both repression and lack of popular support. A massacre of workers in a Siberian goldfield in April 1912, however, set off a spate of strikes and breathed new life into the worker movement, though labor was on the march in the major Western countries as well.

Contemporary observers were divided over Russia's future. Edmond Théry (1854–1925), a well-regarded French economist and prolific author, predicted that "if developments in the major European countries continue from 1912 to 1950 as they did from 1900 to 1912: toward the middle of the present century, Russia will dominate Europe politically, economically, and financially."[1] Others were far less optimistic, but in July 1914 Russia became embroiled in a grueling world war against the greatest land power in the world—imperial Germany—and its allies. Russia mobilized 12 million soldiers. During the course of the war 1,700,000 were killed, 4,950,000 were wounded, and 2,500,000 were taken prisoner. Inflation was rampant. Munition shortages at the start of the conflict hindered the war effort, and bread shortages toward the end sparked popular unrest.

In February 1917, crowds took to the streets in Petrograd, troops mutinied, and the tsar abdicated. For the next several months, power and decision making were divided between liberals and socialists.

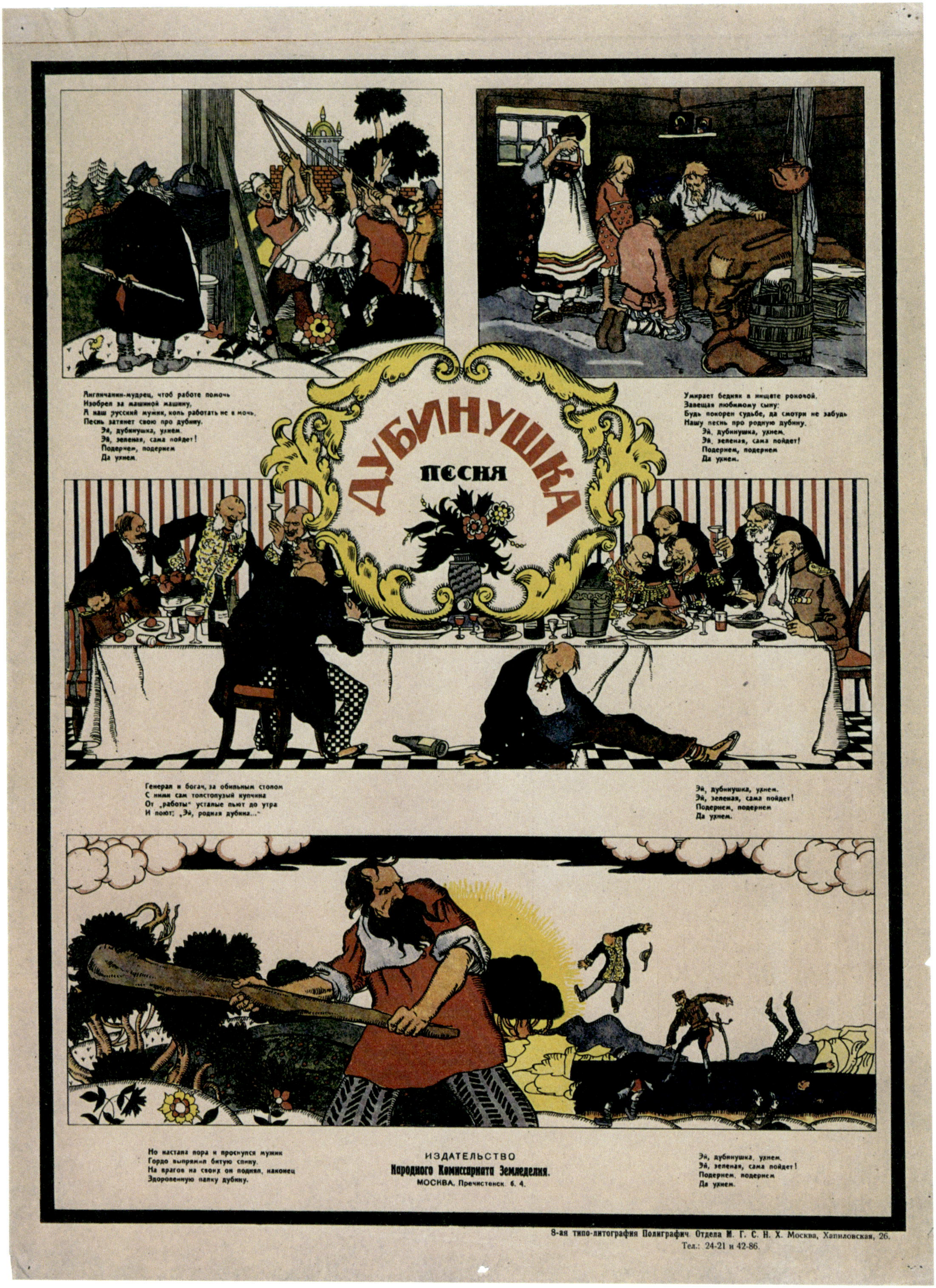
Дубинушка
Песня
Англичанин-мудрец, чтоб работе помочь
Изобрел за машиной машину,
А наш русский мужик, коль работать не в мочь,
Песнь затянет свою про дубину.
Эй, дубинушка, ухнем.
Эй, зеленая, сама пойдет!
Подернем, подернем
Да ухнем.
Умирает бедняк в нищете роковой,
Завещая любимому сыну:
Будь покорен судьбе, да смотри не забудь
Нашу песнь про родную дубину.
Эй, дубинушка, ухнем.
Эй, зеленая, сама пойдет!
Подернем, подернем
Да ухнем.
Генерал и богач, за обильным столом
С ними сам толстопузый купчина
От „работы“ усталые пьют до утра
И поют: „Эй, родная дубина...“
Эй, дубинушка, ухнем.
Эй, зеленая, сама пойдет!
Подернем, подернем
Да ухнем.
Но настала пора и проснулся мужик
Гордо выпрямил битую спину.
На врагов на своих он поднял, наконец
Здоровенную палку дубину.
Издательство
Народного Комиссариата Земледелия.
Москва, Пречистенск. б. 4.
Эй, дубинушка, ухнем.
Эй, зеленая, сама пойдет!
Подернем, подернем
Да ухнем.
8-ая типо-литография Полиграфич. Отдела М. Г. С. Н. Х. Москва, Хапиловская, 26.
Тел.: 24-21 и 42-86.

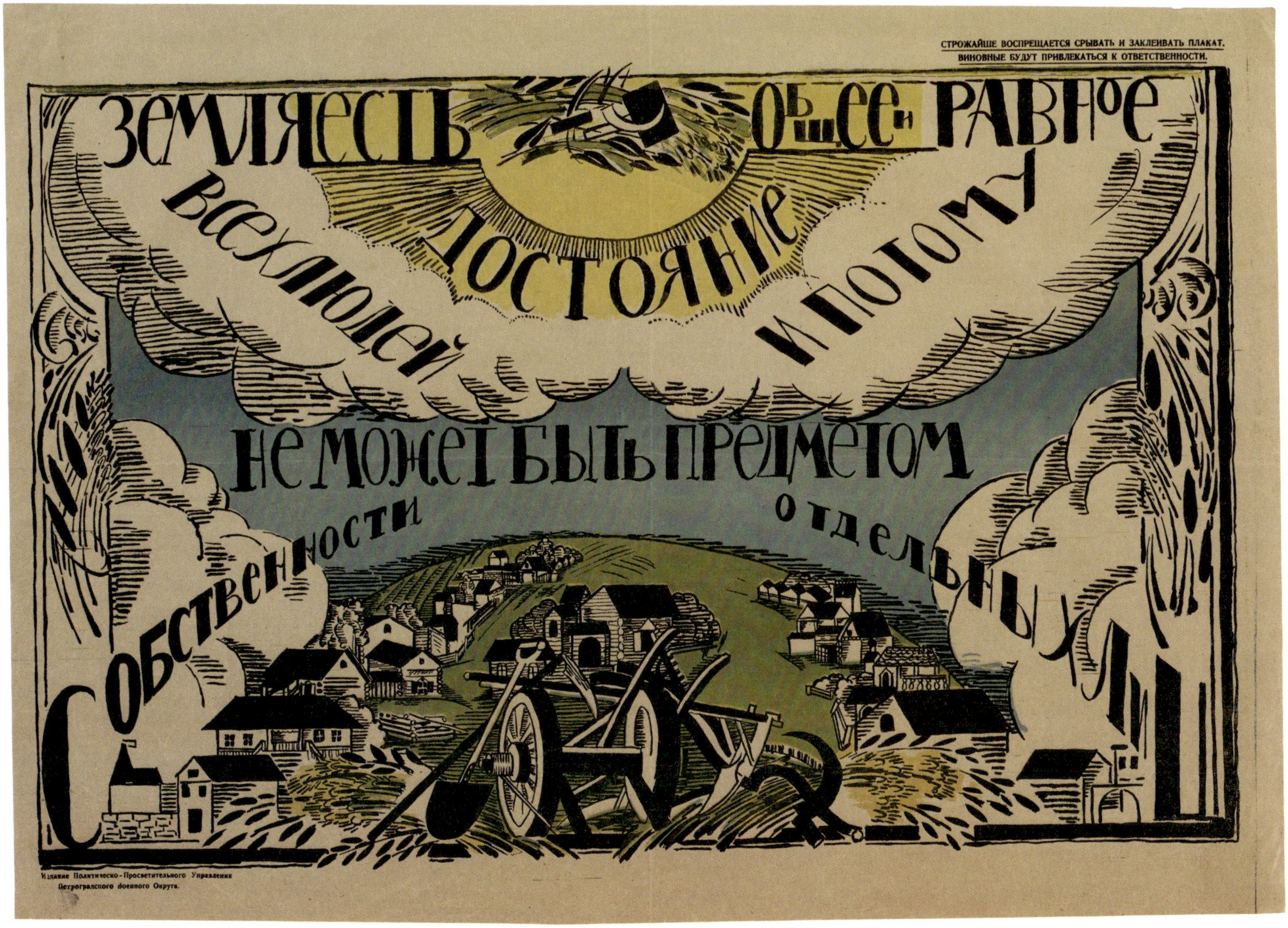

◀ **Fig. 1** V.V. Spassky. "Dubinushka. The song." 1919. The revolutionary song "Dubinushka" expressed the age-old motif of the masses one day rising up to overthrow their oppressors. The Bolsheviks kept the theme alive into the 1930s, well after the traditional Russian elites no longer exploited anyone.

▲ **Fig. 2** Artist not identified. "The land is the common and equal property of all people and therefore cannot be subject to ownership by individual persons." 1917–24?

Crime grew rife. Peasants seized and divvied up many large landholdings. Inflation frittered away the value of the ruble. Popular discontent sharpened.

In October, the Bolsheviks seized power in the name of worker and soldier councils, or soviets. Their first decrees were mostly popular, such as proclaiming an end to the war, approving peasant seizures of land, establishing an eight-hour working day, and granting self-determination to national minorities. Yet the Bolsheviks soon created a one-party dictatorship and a secret police with expansive powers (the Cheka). They also nationalized all banks, other large businesses, financial securities, and urban real estate. In March 1918, they had bought Russia's exit from the war in exchange for one-third of its European territory, which was handed over to Germany (but Russia got most of it back when the war ended in November). In May, committees of the poor were tasked with seizing grain from the peasantry. Inevitably, a civil war broke out, with the Bolsheviks against most of the country. During the civil war, trading grain and other foodstuffs, instead of handing them over to the government at low fixed prices, became a crime punishable by death. Free enterprise and buying and selling goods on the free market were also outlawed. Such policies were known as "war communism."

Thanks to superior military organization, unified political and military leadership, control of the country's heartland and major lines of communication and transportation, a more compelling vision of the future, and the fear of most peasants that the anti-Bolshevik coalition might return private estates to the landlords, by early 1920 the Bolsheviks had won the civil war.

The country, however, lay in ruins. Industrial output had fallen to one-fifth of the prewar level, steel and iron manufacture to 3 percent. Major cities had lost half or more of the population. Livestock production had collapsed. The number of draft animals had fallen by a third. Peasant uprisings multiplied across the country. In early 1921, factory workers struck in Petrograd; sailors at the nearby fortress of Kronstadt rebelled. The Communist authorities supplied the workers with more food and clothing and crushed the rebellion with heavy firepower. Lenin also convinced his colleagues to "retreat on the economic front" by stopping grain confiscations and decriminalizing the retail trade in agricultural and other products. The retreat was to be temporary, because for Marxists not only private property but also free economic exchange had to be eradicated to build socialism.

Peasant uprisings continued into 1922, but again the rebels were no match for a battle-hardened army. The government also systematically disarmed the peasantry. Tragically, the widespread confiscation of grain, which had depressed planting, was exacerbated by an extreme dry spell. A horrific famine resulted. By the end of 1922, some 5 million people had died from starvation and disease. The government, unlike in 1891, provided little support to the hunger victims, nor did it allow civic organizations to help, except foreign agencies like the American Relief Administration. This assistance was decisive, however: without it, millions more would have died.

Sporadic but very serious hunger continued into 1924, yet big grain exports recommenced in 1923. Peasant unrest persisted. A massive rebellion broke out in Georgia in late August 1924. It was crushed, leaving many thousands dead. The repression was characteristically accompanied by religious persecution, since the Bolsheviks viewed "counterrevolutionary" peasants, capitalists, and the religious as inextricably bound

Fig. 3 Artist not identified. "Petr and Vasilii, or the village in «Sovdepiia»." [1919]. The Russian Civil War (anti-Bolshevik) poster recounting the story of a committee of the poor that ruins the village. Published in Odessa by Odesskoe Otdelenie Otdela Propagandy.

Петръ и Василій или деревня въ „Совдепіи“

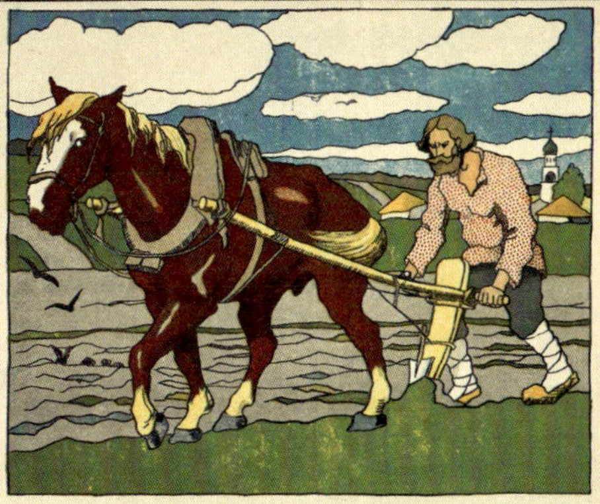

1. Петръ трудился надъ своей нивой.

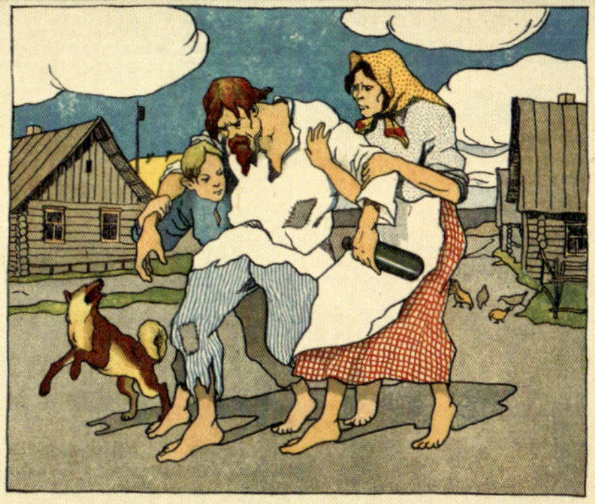

2. Василій же былъ и день и ночь пьянъ.

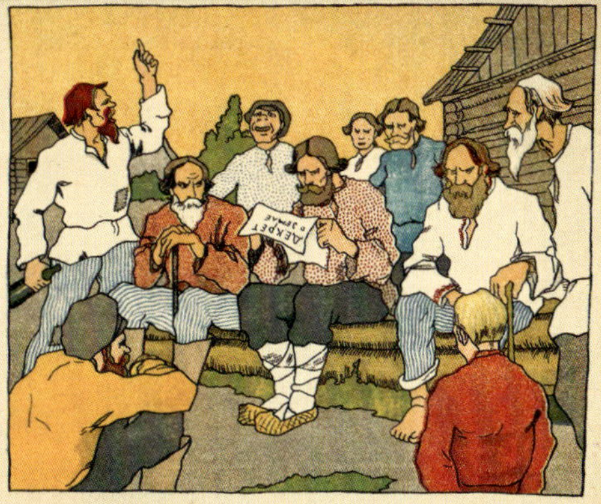

3. Прислали московскіе комиссары декретъ о землѣ.

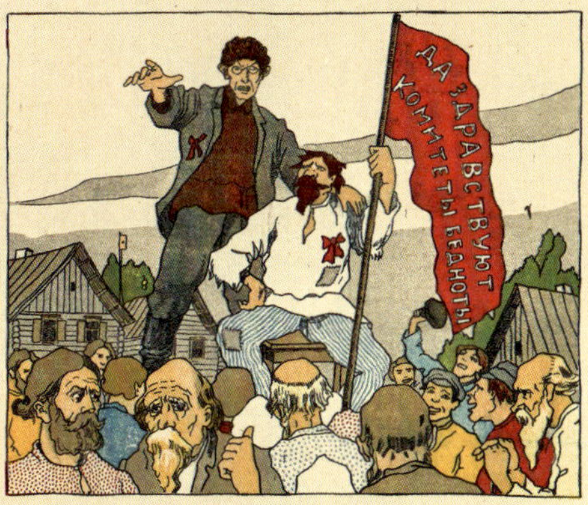

4. Пріѣхалъ ораторъ и Васька сталъ предсѣдателемъ комитета бѣдноты.

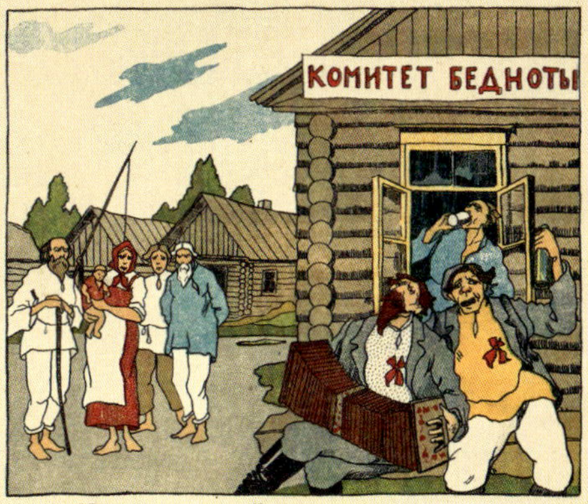

5. И пошла для Васьки сладкая жизнь.

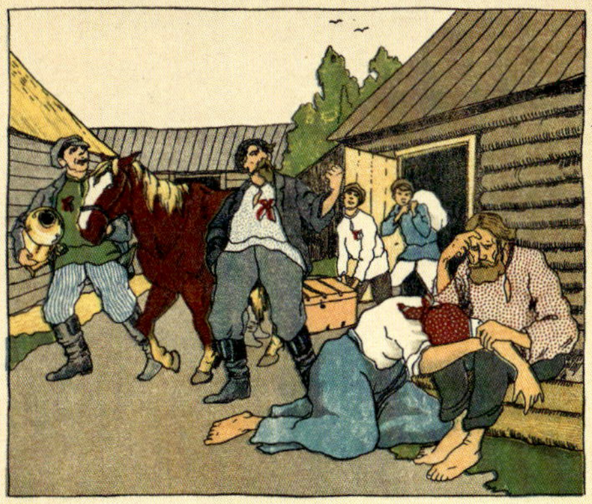

6. Кто вѣкъ трудился—вреднымъ кулакомъ объявленъ.

7. И вотъ что, спустя мѣсяцъ, стало съ деревней.

1. Въ одной деревнѣ жили два крестьянина, Петръ и Василій.

Петръ былъ мужикъ работящій, честно трудился надъ своей землей, и домъ у него былъ полная чаша.

2. Василія все село презрительно звало Васькой. День и ночь пьяный, вѣчно съ бутылкою въ рукахъ, онъ былъ горемъ своей семьи.

3. Свалилось на Святую Русь большое несчастье, пришла совѣтская власть.

Въ деревню московскіе комиссары прислали декретъ о землѣ. Прочитали и призадумались старики. Недоброе почуяли..... А Васька понялъ, что на его пьяную улицу пришелъ большой праздникъ...

4. И когда отъ Ленина пріѣхалъ ораторъ, то пьяница и горлодранъ Васька оралъ на митингѣ больше всѣхъ и первымъ получилъ власть надъ деревней. Онъ пріосанился, нацѣпилъ красный бантъ, руки въ боки....

Теперь онъ—самый главный предсѣдатель комитета бѣдноты, и уже Васькой его не назови,—сейчасъ же разстрѣляетъ!

Зови его Василіемъ Ивановичемъ!

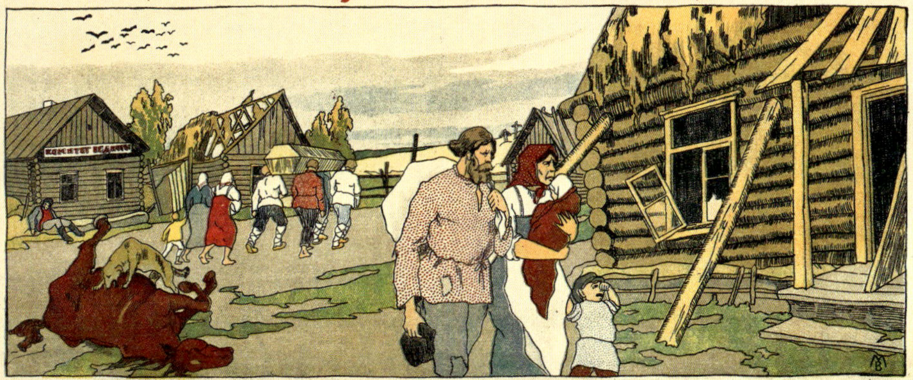

5. И пошла для Васьки сладкая жизнь, вѣчный праздникъ!... Кто трудился весь свой вѣкъ, тотъ босой ходитъ. А Васька въ лакированныхъ сапогахъ, кишмешевку ведрами пьетъ, на гармоніи революціонныя пѣсни играетъ.... Не жизнь, а малина! Умирать не надо!

6. Дальше больше,-началъ Васька все строже порядки въ деревнѣ наводить, объявивъ соціализацію имущества. Свелъ у Петра лошадь, утащилъ послѣднюю муку, унесъ самоваръ, сундукъ съ добромъ, которое дѣды и прадѣды наживали.

Вышло по декрету такъ: кто работалъ всю жизнь, не покладая рукъ, кто трудился-тотъ кулакомъ и вреднымъ человѣкомъ объявленъ.

7. И вотъ, что, спустя мѣсяцъ, стало съ деревней.

Кто по міру съ дѣтишками пошелъ, а кто и въ сыру-землю отправился. Передохла скотина, и сыты стали только одни комитеты да собаки....

Такъ науправлялъ на Руси большевикъ Васька.

together in opposition to building socialism. Spreading knowledge of the events worried peasants throughout the country. Heavy taxation, confiscation of property for failure to pay, and the high price of manufactured goods led to ongoing discontent in the countryside, which sometimes expressed itself in violent attacks against officials and rural correspondents of the state-controlled press (who many peasants believed were spies). In August 1924, for the first time, a key secret police summary report contained the rubric "kulak terror," meaning violent attacks by well-to-do peasants against authority figures. (The Bolsheviks' Marxist outlook required them to divide the population into social classes. In reality, there was very little social differentiation in the Russian countryside after the spontaneous land reform of 1917.)

All these problems in the villages led the Communist authorities at their October 1924 plenum of the Central Committee to enunciate a "face to the village" campaign. For the next year, the party paid attention to rural problems and strived to mend fences with the peasantry. At a Politburo meeting in January 1925, Mikhail Kalinin (1875–1946) pointed out that the typical "kulak"[2] lived less well than the typical worker. In an April 1925 speech, Nikolai Bukharin (1888–1938),[3] a leading Politburo member, denounced the fear of autonomous economic growth in the countryside, where hardworking peasants were afraid to buy farm implements because it would get them labeled "kulaks." On the contrary, he asserted, "enrich yourselves, accumulate, develop your farms."[4] (Even at the time he was roundly criticized for such a "pro-capitalist" assertion; later it helped cause his doom.) Thousands of party members were sent to the countryside to work with the peasants and even to learn from them, but neither the helpers nor helpees enjoyed this collaboration very much. In fact, the city folk, who were already deeply suspicious of the peasantry, were often horrified by what they saw: political ignorance, illiteracy, deep suspicion of the urban world, resentment of the higher standard of living and greater privileges of industrial workers, brawling, disinterest in building socialism, efforts to co-opt rural Soviets for their own ends, intermarriage of government officials and "kulaks," drunkenness, and corruption. Many local officials, who were ill disposed toward rural folk to start with, continued to resort to coercive methods. Since grain collections did not increase, the policy was abandoned in late 1925.

Lenin's New Economic Policy (NEP) nevertheless worked like magic. By 1926, industrial production slightly exceeded prewar levels (of course, the European countries had advanced vastly farther), and more land was being cultivated, though the grain harvest still fell short (despite continued population growth). Grain was important: 80 percent of all arable land was sown to grain, and it was easy to export. The state dominated heavy industry and large-scale manufacturing, but private entrepreneurs controlled handicraft production, much of the retail sector, and nearly the whole of agriculture. During the revolution, the peasantry had seized all of the agricultural land and had forced back into the communes those peasants who had set out on their own starting in 1906. As a result, the agricultural sector was less efficient, more old-fashioned in its methods, and less oriented toward the long-distance market than before 1917—but still vastly more productive than during war communism.

Lenin and other Bolshevik leaders hoped to draw the peasantry into the cooperative sector, where they would join together to buy, sell, provide credit, or produce food,

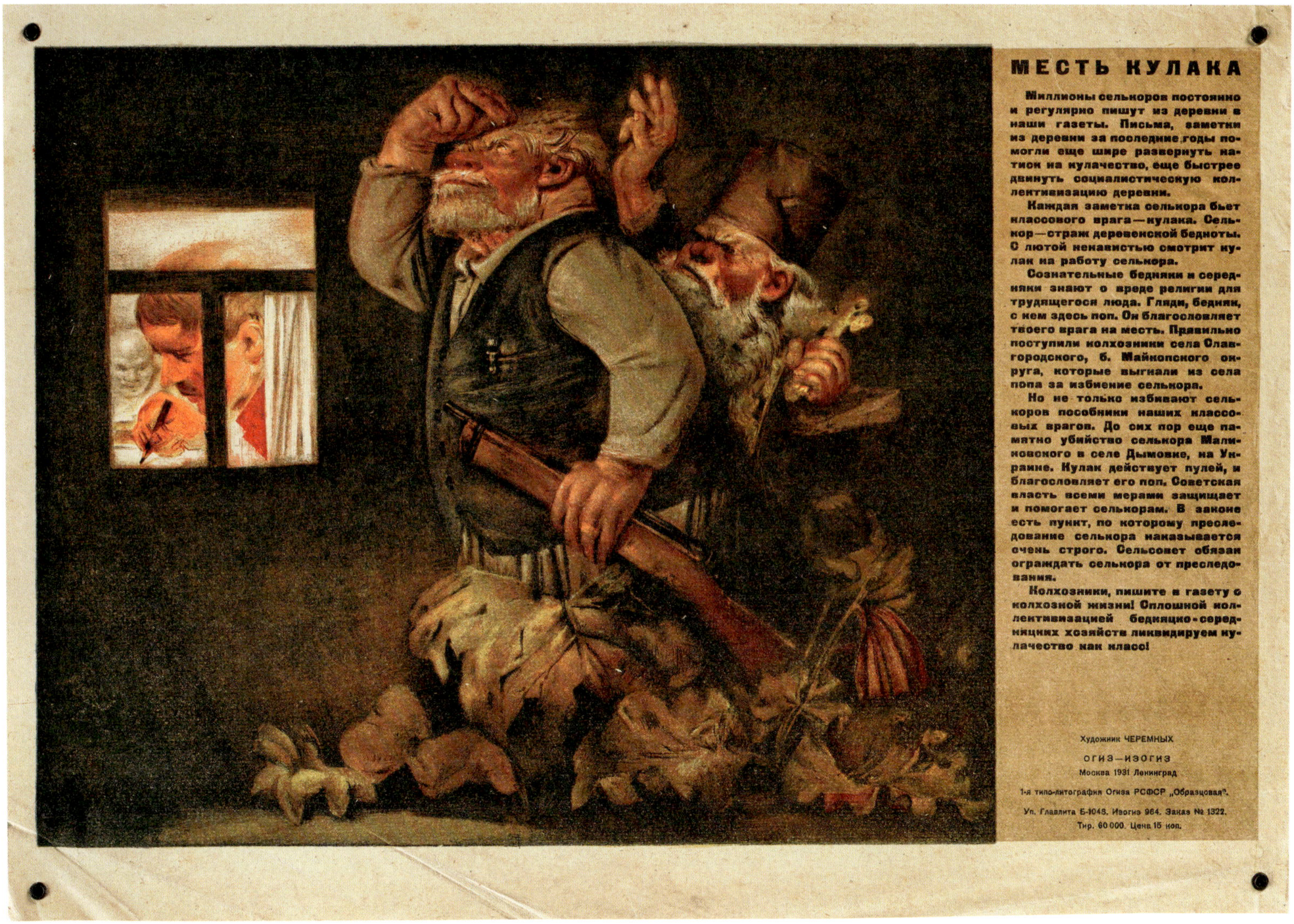

Fig. 4 M.M. Cheremnykh. "The revenge of the kulak." 1931. A priest inciting a "kulak" to murder a rural correspondent.

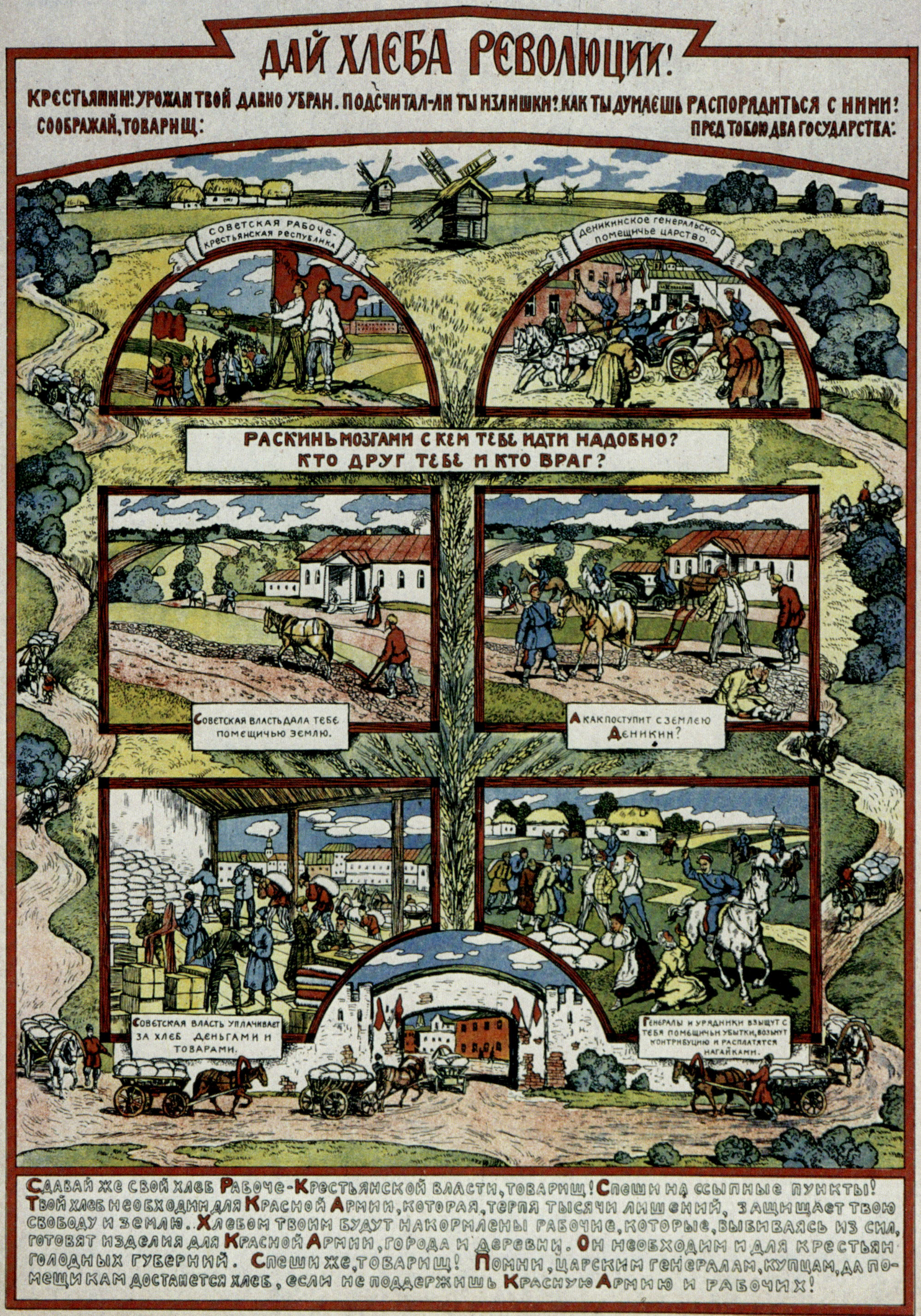
ДАЙ ХЛЕБА РЕВОЛЮЦИИ!
КРЕСТЬЯНИН! УРОЖАЙ ТВОЙ ДАВНО УБРАН. ПОДСЧИТАЛ-ЛИ ТЫ ИЗЛИШКИ? КАК ТЫ ДУМАЕШЬ РАСПОРЯДИТЬСЯ С НИМИ?
СООБРАЖАЙ, ТОВАРИЩ:
ПРЕД ТОБОЮ ДВА ГОСУДАРСТВА:
Советская Рабоче-Крестьянская Республика
Деникинское генеральско-помещичье царство.
РАСКИНЬ МОЗГАМИ С КЕМ ТЕБЕ ИДТИ НАДОБНО? КТО ДРУГ ТЕБЕ И КТО ВРАГ?
Советская власть дала тебе помещичью землю.
А как поступит с землею Деникин?
Советская власть уплачивает за хлеб деньгами и товарами.
Генералы и урядники взыщут с тебя помещичьи убытки, возьмут контрибуцию и расплатятся нагайками.
Сдавай же свой хлеб Рабоче-Крестьянской власти, товарищ! Спеши на ссыпные пункты! Твой хлеб необходим для Красной Армии, которая, терпя тысячи лишений, защищает твою свободу и землю. Хлебом твоим будут накормлены рабочие, которые, выбиваясь из сил, готовят изделия для Красной Армии, города и деревни. Он необходим и для крестьян голодных губерний. Спеши же, товарищ! Помни, царским генералам, купцам, да помещикам достанется хлеб, если не поддержишь Красную Армию и рабочих!
Государственное Издательство
Москва — 1920 г.

◀ **Fig. 5** K.V. Spassky. "Give bread to the revolution!" 1920. "Peasant! Your harvest has long been gathered. . . . Give up your grain to the Worker-Peasant Authority, comrade!"

both to create economies of scale and to wean the peasantry away from the evils of the market. After all, according to Marxist doctrine, peasants were by definition petty bourgeois and indeed enemies of the proletariat. The cooperatives, Lenin hoped, would gradually suppress private trade, defend poor and middle peasants against the "kulaks," serve as a "school of collective economics," overcome the individualistic character of the peasantry, and expand socialist construction in the village. By 1927, half of all peasants made use of consumer cooperatives. Yet most peasants remained skeptical of producer and credit cooperatives. The better-off members tended to dominate them, few cooperatives achieved a high level of efficiency, many seemed to have a political agenda, and they were gradually integrated into the state apparatus. Many Bolsheviks also looked askance at them, for they remained a form of private ownership. Lenin himself had vacillated about whether cooperatives tended more to foster the development of capitalism or socialism. All the Bolsheviks

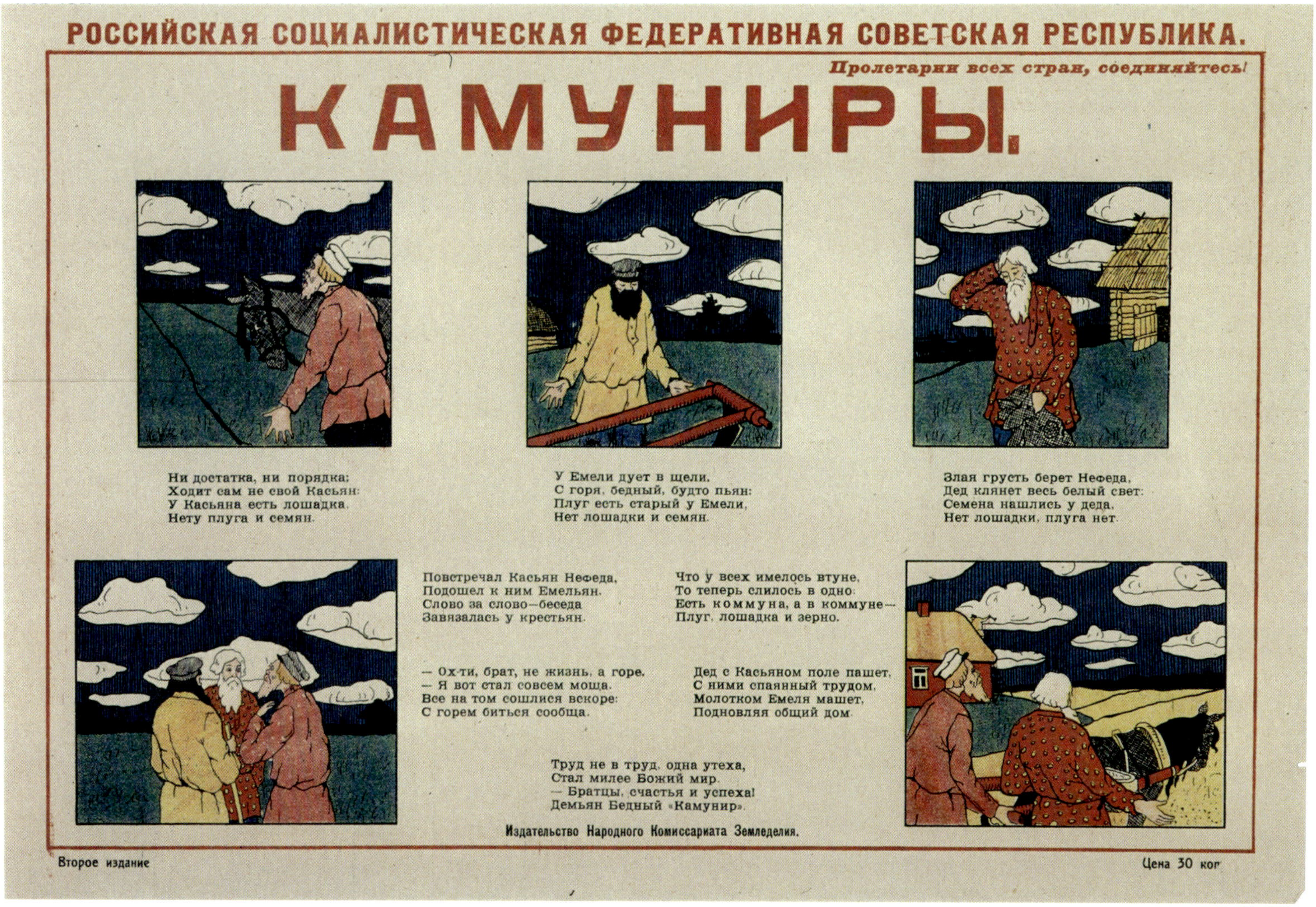

▼ **Fig. 6** Artist not identified. "Kamuniry," 2nd edition. 1919. Poem "Communards" by Demian Bednyi. The benefits of cooperation: Kasian has a horse but neither plow nor seeds. Emelian has a plow. Nefed has seeds.

looked to the eventual creation of a colossal network of big, centrally controlled, highly mechanized socialist farms. They simply disagreed about how soon.

Leaders compensated for the "economic retreat" by clamping down on the "political front." (Leading Bolsheviks like Lenin viewed the struggle for socialism in militaristic terms.) The same party congress that adopted the NEP also imposed a ban on factions within the Communist Party. Not only were no other political parties allowed to take part in government, but even within the ruling Communist Party it became illegal to form coalitions or groupings. Leaders of other political parties, as well as of organizations with a potential for political influence, like the Russian Orthodox Church, suffered arbitrary arrest, trial, and execution, as well as other forms of persecution.

Despite official commitment to the *smychka*, the officially proclaimed union of peasants and workers, most peasants remained deeply mistrustful of the Communist authorities, who for their part saw the peasants as ignorant and superstitious. Officials who came to the villages typically had no idea about farming, but one still had to listen to them and even implement their orders. They tended also to be nicely dressed and far better-off than the villagers. Some locals referred to them as "nomads."[5] The best disposed aimed to discipline, civilize, and Sovietize their charges; the worst, to swindle, cheat, abuse, and lord it over them. Many drank plentifully and caroused. Given their low salaries, most saw bribes as a basic form of compensation. All the peasants really wanted was economic freedom, reasonable taxes, fair prices, and the right to benefit from the fruits of their labor. They were willing to follow all the rules and abide by the law, so long as this did not ruin them financially, which in many cases it unfortunately did. Perhaps surprisingly, many peasants had more faith in the party leaders in faraway Moscow, whom they considered "ours" and believed were trying to "bring about socialism, that is the reign of God on earth."[6] Still, few peasants had the intellectual ability to formulate a political position, much less a program. Although one can think of the NEP as a sort of "golden age of the Russian peasantry," most peasants during this era were quite unenthusiastic about the present and pessimistic about the future. The party leaders, for their part, were pursuing contradictory goals. On the one hand, they wanted agricultural development, but on the other they were afraid of stronger peasant households and the "class warfare" this would bring.

Lenin had suffered a series of strokes, leaving him debilitated from early 1923 and killing him in January 1924. A struggle for power had already begun. Joseph Stalin (1878–1953) had many disabilities: he was a Georgian who spoke Russian with an accent, was short and had a pockmarked face, lacked the intellectual talents of other Bolshevik leaders, and was a poor orator. He was nonetheless in an especially strong position. As general secretary of the Central Committee of the Communist Party, he could divvy out to his supporters the exponentially increasing patronage jobs controlled by the party. Until Leon Trotsky (1879–1940) was expelled from the Soviet Union in early 1929, Trotsky's charisma, prestige, rhetorical skill, and intellectual brilliance made him the most obvious potential dictator and therefore the greatest threat to the other leaders. By contrast, Stalin feigned humility and presented himself as a team player, eager to defer to his colleagues as a variety of coalitions formed, first against the "left deviation" of Trotsky and then against the "right deviation" of Bukharin.

Trotsky and his supporters advocated rapid industrialization. Building up industry required huge investments. Yet all obvious sources of wealth had already been expropriated, including the church, the nobility, the merchantry, the state gold reserves, businesses, and most personal riches. Moreover, the former government's massive debts to the leading European countries had been repudiated. Who would lend them more? The peasantry was really the only major source of wealth in the country, and certainly the government desperately needed to obtain a portion of their agricultural output. But to do so, it needed manufactured products to exchange, which brought one right back to the problem of industrial weakness. So Trotsky advocated squeezing the peasantry. The party majority during the first years of the NEP, whose chief spokesman was Bukharin, argued for maintaining Lenin's policy of *smychka* (alliance). If the peasants prospered, it was asserted, they could pay more taxes, purchase more industrial goods, and help grow the economy without coercion or other mistreatment. Since the peasantry made up 85 percent of the population, antagonizing it would be political suicide. Stalin supported this position. But he also began, as early as 1924, to set out in his own direction, revising Leninism significantly. Back in fall 1923, a Moscow-backed insurrection in Germany had failed, dealing a crushing blow to both leading and rank-and-file activists in the USSR. Stalin now argued that, despite Russia's backwardness, its industrial workers, or proletariat, could "build socialism in one country," even if no further socialist revolutions occurred in the leading European countries—the only hope for the success of the Russian Revolution, according to Lenin, Trotsky, and other leading Bolsheviks. It is not hard to see why Stalin's new theory was popular among ordinary Russians.

Once Trotsky had been defeated, and in 1927 expelled from the Communist Party, Stalin began to inch toward Trotsky's economic position. He commissioned a study purporting to show, with a fudging of statistics, that only half as much grain had been brought to market in 1927 as in 1913. The supposed explanation was the loss of big, efficient estates to peasant land seizures. The Bolshevik answer was not, of course, to return the property to big landowners, but instead to gather the peasants into larger communal organizations—into big, modern, technologically advanced farms—to collectivize them. The trouble was, few Russian peasants wished to give up their autonomy and hand over their land allotments and livestock to an anonymous collective, despite various incentives provided by the state. (Nor was there the industrial capacity to produce sufficient farm machinery.) So the party leadership hiked the taxes on the peasantry (especially the better-off), slashed the official price paid for grain, excluded private traders from grain procurement, and raised the cost of manufactured goods. The purpose of all of these policies was to extract wealth from the peasantry, though in a gradual fashion. Already beginning in August 1926, the secret police had received orders to facilitate grain deliveries. They also arrested thousands of private traders and "kulaks." Many peasants responded by withholding their produce, planting other crops, failing to pay their taxes, and refusing to buy factory-made merchandise. In 1927 and 1928, the government felt the pinch, the more so as it had planned to increase industrial development, for which it needed grain exports to trade for imports of machinery. In fact, the situation was so bad that on March 11, 1927, strict orders were given to all publications within the state-controlled information system (essentially

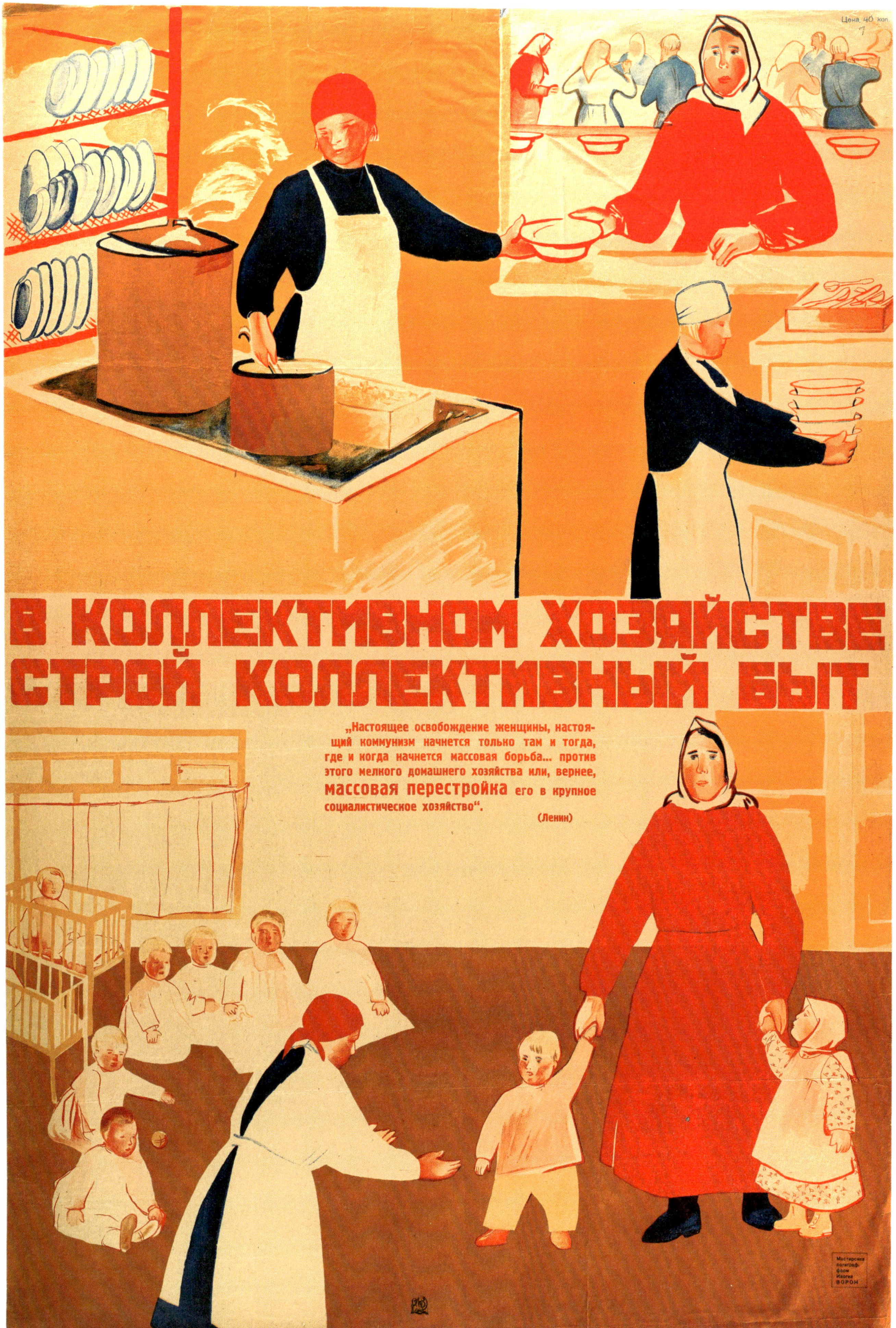
Цена 40 коп.
В КОЛЛЕКТИВНОМ ХОЗЯЙСТВЕ
СТРОЙ КОЛЛЕКТИВНЫЙ БЫТ
„Настоящее освобождение женщины, настоящий коммунизм начнется только там и тогда, где и когда начнется массовая борьба... против этого мелкого домашнего хозяйства или, вернее, массовая перестройка его в крупное социалистическое хозяйство“.
(Ленин)
Мастерская полиграф-форм Изогиз ВОРОН
ИЗОГИЗ
Москва 1931 Ленинград
Тираж 20 000

all the country's media) forbidding any mention of problems with grain supply.

A number of factors came to Stalin's aid. From 1926, the Soviet leadership worried more and more about rural crime, waves of arson, violent attacks against officials, hooliganism, "banditry," growing "kulak" influence, and a unified front within the village against party and state. Indeed, fewer and fewer Communists—and more and more "kulaks"—were elected to local office in the countryside. The closer officials looked at the Russian countryside, the more it appeared violent, primitive, menacing, and lawless. In many regions, criminal gangs made it impossible to travel from village to village in safety. The authorities feared the prestige of bandits and criminals in general. Even Bukharin, the great defender of the peasants, repeatedly called for an offensive against the "kulaks," for example, in January 1926 and October 1927.

In reality, few peasants could be considered wealthy. An official survey of 1927 found that 3.2 percent of households owned on average 2.3 draft animals and 2.5 cows. These were the rich. By contrast, 80 percent of households possessed no draft animals at all, and 57 percent had no cow.[7] Only 1.4 percent of peasant households had more than 43 acres of sown cropland, 3.7 percent had 27 acres or more, and 36.7 percent less than 5.5 acres, whereas the *average* farm in the United States had nearly 200 acres.[8] Yet the Marxist concept of class conflict required the Bolsheviks to dream up a rural bourgeoisie. The handiest marker in this regard seemed to be hiring labor. Yet only 1 percent of rural households hired more than one worker, most worked only part-time, many families with a lot of mouths to feed but few work-capable members hired labor, and better-off households often avoided hiring labor for fear of being labeled "kulaks." Indeed, secret police reports repeatedly noted that most peasants saw not class stratification but work stratification, with the lazy and the stupid at the bottom of the village hierarchy.

International tensions and capitalist "encirclement" of the USSR also troubled the party leadership. The slaughter of Chinese Communists in Shanghai in April 1927, the expulsion of the Soviet trade mission from the United Kingdom on charges of espionage in May, and the assassination of the Soviet ambassador to Poland in June underlined the hostile global environment and suggested that war might be imminent. The party leadership used the events as a pretext to beef up the secret police and to conduct "counterrevolutionary" operations. Thousands were searched and arrested around the country but especially in the countryside. The state-controlled media played up the threat of war, leading to panic, hoarding of foodstuffs, and widespread slaughter of livestock. Unemployment soared. Popular discontent swelled. Secret police investigators uncovered pervasive antiwar and even defeatist sentiment in July and August. Orders went out on September 12 to make no mention in the press or public appearances about grain exports. This regime of tight information control and secrecy made it easier to launch a frontal assault against the peasantry when the time came.

Preparations for an attack were in fact under way in fall 1927. In mid-October, the government authorized the secret police (now called the OGPU) to arrest on their own authority both traders engaged in "speculation" and officials who abetted them. On October 29, Genrikh Iagoda (1891–1938), the head of the OGPU, proposed empowering his agency not only to arrest but also to sentence "speculators" administratively. Ironically,

Fig. 7 Artist not identified. "In the collective farm, build a collective lifestyle." 1931. "The genuine liberation of women, genuine communism begins only when and where commences a massive struggle . . . against the small-scale household economy, or rather the massive transformation of it into a big socialist economy. – Lenin."

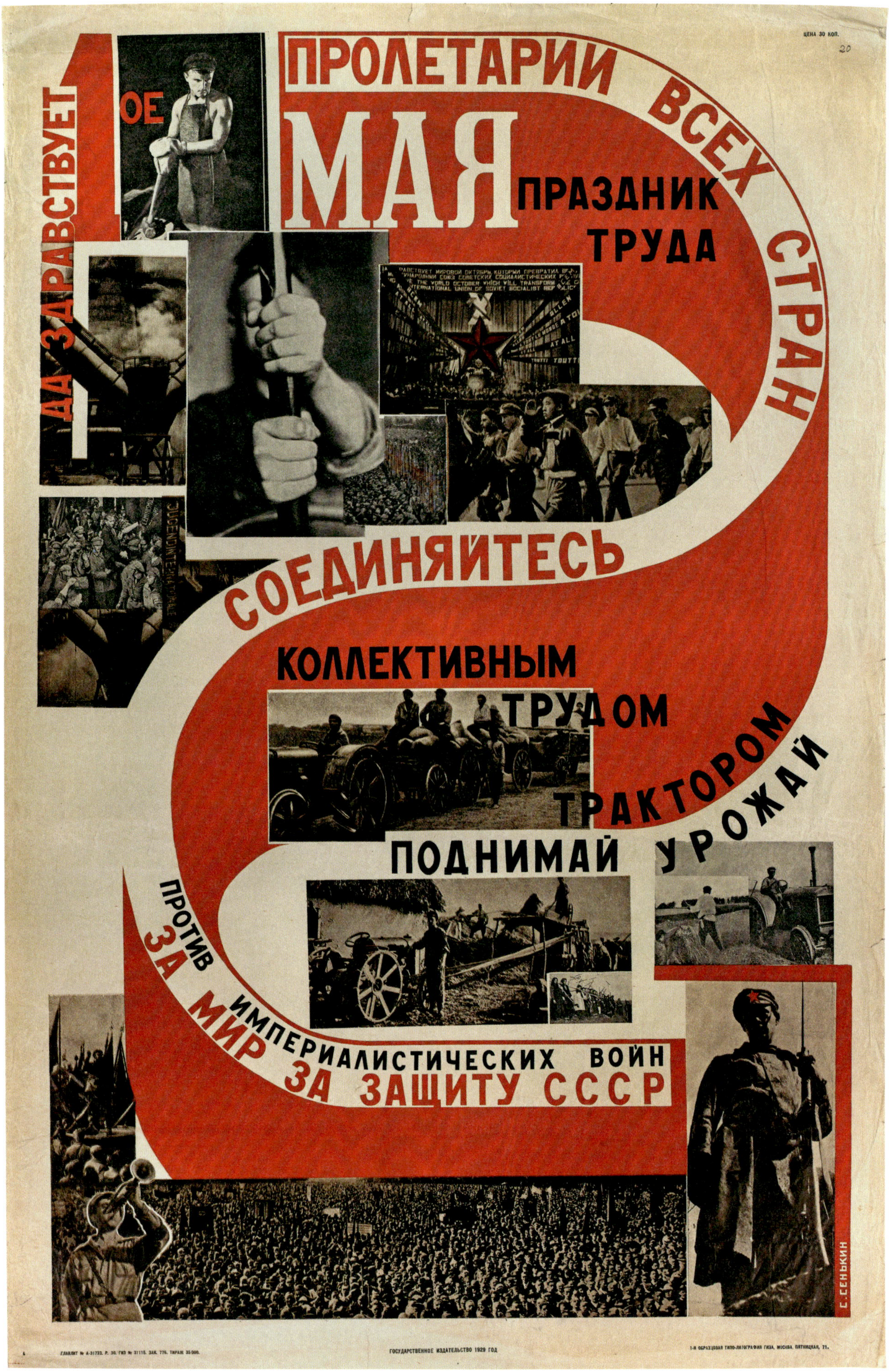
ЦЕНА 30 КОП.
ДА ЗДРАВСТВУЕТ
1 ОЕ
МАЯ
ПРОЛЕТАРИИ ВСЕХ СТРАН
ПРАЗДНИК
ТРУДА
СОЕДИНЯЙТЕСЬ
КОЛЛЕКТИВНЫМ
ТРУДОМ
ТРАКТОРОМ
ПОДНИМАЙ УРОЖАЙ
ПРОТИВ
ИМПЕРИАЛИСТИЧЕСКИХ ВОЙН
ЗА МИР ЗА ЗАЩИТУ СССР
С. СЕНЬКИН
ГЛАВЛИТ № А-31733. Р. 30. ГИЗ № 31115. ЗАК. 776. ТИРАЖ 35 000.
ГОСУДАРСТВЕННОЕ ИЗДАТЕЛЬСТВО 1929 ГОД
1-Я ОБРАЗЦОВАЯ ТИПО-ЛИТОГРАФИЯ ГИЗА, МОСКВА, ПЯТНИЦКАЯ, 71.

Fig. 8 S.Ya. Senkin. "Long live May 1, the celebration of labor." 1929. "Proletarians of all countries, unite! Against imperialist wars. For peace. For the defense of the USSR. Increase the harvest with collective labor and the tractor."

Stalin rejected the proposal as going too far. It seems hard to imagine he was not feigning moderation.

The announcement, during the celebration of the tenth anniversary of the October Revolution a couple of weeks later, of a seven-hour workday for factory workers outraged peasants, who already considered urban workers both lazy and coddled by the state. By December, the secret police was warning of a mounting confrontational atmosphere among the peasantry. In mid-December, at the Fifteenth Party Congress, which expelled Stalin's former allies, Lev Kamenev (1883–1936) and Grigory Zinoviev (1883–1936), from the party, the leadership made a commitment to the gradual collectivization of agriculture, but the tone was moderate. Stalin himself proclaimed that one had to do away with the "kulaks" only by legal means. "Those comrades," he argued, "are wrong who think that it is possible and necessary to put an end to the kulaks by means of administrative measures, through the GPU: give an order, affix a seal, and that settles it. That is an easy way, but it is far from being effective. The kulak must be defeated by means of economic measures and in conformity with Soviet law. Soviet law, however, is not a mere phrase."[9] It seems unlikely he meant it. More than other party leaders, Stalin viewed the peasants as a largely undifferentiated mass, for the most part hostile to the party, the workers, and the building of socialism. Throughout the fall he warned that "kulaks" were purposely withholding grain and convincing middle peasants to do the same. He also knew that the state grain procurements were way down from the previous year.

Soon after the congress, on December 19, the secret police and other government agencies began demanding a "sharp break in grain collection" and warning of serious consequences for officials who failed to comply. Many of these directives bore Stalin's signature.

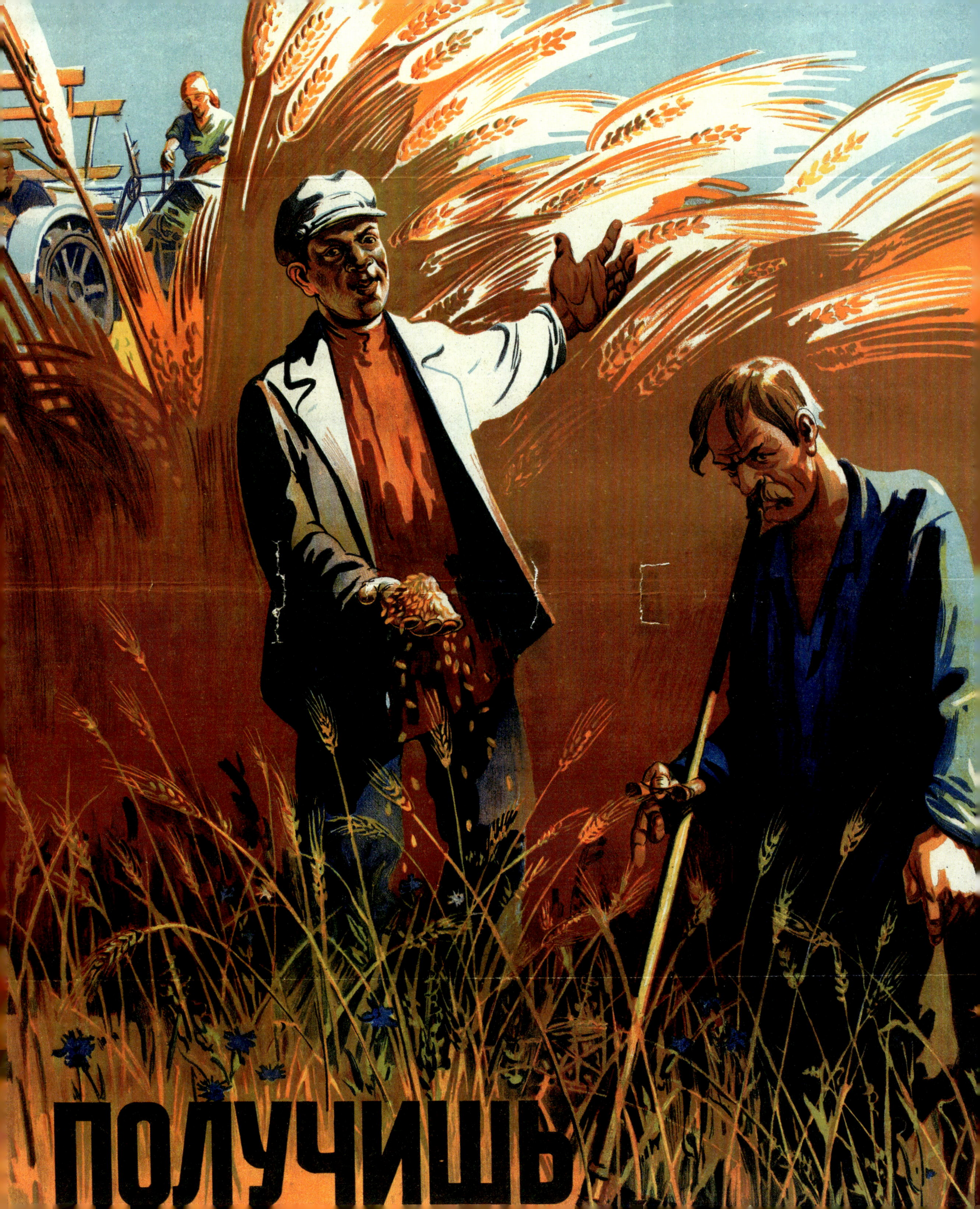
ПОЛУЧИШЬ

Chapter 1

INCHING TOWARD ARMAGEDDON, 1928–1929

Throughout 1928 and 1929, the Communist authorities adopted diverse means to squeeze grain out of the peasantry, short of a wholesale assault and complete takeover of the countryside. The OGPU was at the forefront of these grain-confiscation campaigns. In fact, during these operations the secret police became a central player in Soviet governance, a position it would occupy until the collapse of Communism in 1991 and, with the rise to power of former KGB operative and FSB chief Vladimir Putin, until the present day.

Already on January 4, 1928, the OGPU (secret police) instructed its agents to get directly involved with grain collection as enforcers. They were to act quickly, demonstratively, and extrajudicially. The next day, the Politburo unanimously adopted a directive, signed by Stalin and sent out on January 6 to the party apparatus, demanding immediate mobilization of the entire Communist Party to implement the collection campaign using harsh administrative measures, in particular against "kulaks." The militia and prosecutors should also be drawn into the work, since of course the leading officials of all Soviet institutions were Communist Party members. Failure to comply would be treated as a breach of party discipline. It is not surprising that many officials preferred to overfulfill than to underfulfill such orders. The fact that the secret police acted first in this drama might suggest that it was in charge, but this seems unlikely. Although documentary proof is lacking, one can assume that Stalin collaborated with Genrikh Iagoda, the second in command of the OGPU,[1] behind the scenes and essentially brought him to the center of the operation as an ally.

Iagoda and his henchmen were also unleashed against the political opposition. On January 17, the police seized Trotsky, carried him physically out of his Moscow apartment, and deported him to Alma-Ata, at the far end of Kazakhstan, 4,000 kilometers from Moscow. Stalin gleefully jotted "ha ha" in the margin of the report.[2]

Agents of repression throughout the country immediately went to work. They were typically organized in teams of three, or troikas, representing the secret police, prosecution, and party. They arrested "speculators," "kulaks," and others who got in the way of the collection campaign.

In addition to a steady stream of directives, the Politburo also decided on January 9 to send top party officials to the chief grain-producing regions of the country to force the issue. Stalin himself departed for Siberia on January 14, having signed a party directive asserting that attacking "kulaks" was the best way to convince "middle peasants" to hand over grain at low state prices. The order concluded with a typical Bolshevik flourish: grain procurements were a "fortress we must capture no matter what."[3] Stalin's closest associates, Viacheslav Molotov (1890–1986) and Anastas Mikoian (1895–1978), were sent to the Ural

◂ Detail of "Sow high-yield seeds," *see page 28.*

Mountains and North Caucasus regions, respectively, and with the same purpose.

Even before Stalin's departure, the OGPU and the chief prosecutor in Siberia had directed their personnel to punish peasants withholding large quantities of grain by means of article 107 of the criminal code, an obscure and little-used statute banning speculation. When he arrived in Novosibirsk five days later, Stalin claimed that prosecution under article 107 was yielding excellent results in other regions. (He may have been bluffing.) He traveled across Siberia, cajoling and threatening officials not to "wimp out," demanding "savage" pressure,[4] repeatedly raising the collection targets, and insisting on the sacking of officials complicit in grain "speculation." A significant purge of less-forceful party members in fact ensued. After Molotov departed from the Urals, for example, 1,157 such officials were removed from office, excluded from the party, or brought to trial.[5]

The day Stalin headed back to Moscow, February 4, the secret police ordered its operatives to telegraph the results of the campaign. The repression had paid off. Twice as much grain was collected in January 1928 as in December 1927, yet the takings fell short of the year before (only 300 compared to 428 million poods). Perhaps for this reason, the party leaders did not crow. The campaign in fact had been waged in total secrecy. Not a word was breathed in the press—not even the mention of "kulaks" or article 107. On January 26, 1928, the party mouthpiece *Pravda* had even denounced rumors about the end of the NEP, a return to civil war era confiscations, and impeding famine. Of course, Moscow could not prevent a hoarding mentality from taking hold, given the war hysteria it had stirred up. A police report on February 11 quoted one well-to-do peasant in Voronezh province saying: "We can live without manufactured cloth and can use home-woven cloth; but we need to conserve grain, which will be necessary in the event of war."[6] In any event, the peasants were unlikely to indefinitely hand over grain at low prices and would soon cut their output. Indeed, complaints came in from the villages that not enough was left for survival until the next harvest. Yet the urban population was growing and needed more and more grain. Something had to give.

On February 13, a party directive signed by Stalin admitted that the grain crisis had not ended. Workers and urban dwellers would face shortages and wage cuts. Heavier taxes on "kulaks" were needed. Yet officials up and down the apparatus knew full well that the vast majority of peasants holding back grain in the hope of higher prices were middle peasants. Repression continued. During the first two months of the year, 6,854 people had been arrested for obstructing the collection campaign.[7] The hardliners could well exult: grain collection was up 186 percent in February from the previous year, and 50 percent over January. Yet they wanted more, and the campaign continued. On March 21, *Pravda* warned that some rich peasants had gained possession of 300 acres of land. Within two months, some 74,000 acres had been confiscated from 6,214 people, or just over 12 per person—clearly not big landowners.[8]

A second wave of coercive measures commenced in April. Local officials applied such intimidation and dirty tricks as locking peasants up for a day or two, brandishing firearms, waiting for the man of the house to go out, and pressuring the illiterate to sign away their property. Much of the rhetoric was about teaching the peasants a lesson. "We will show you that you live in a revolutionary state of workers and peasants," warned one official quoted in *Pravda* on March 16.[9] Thousands of peasants desperately wrote to newspapers

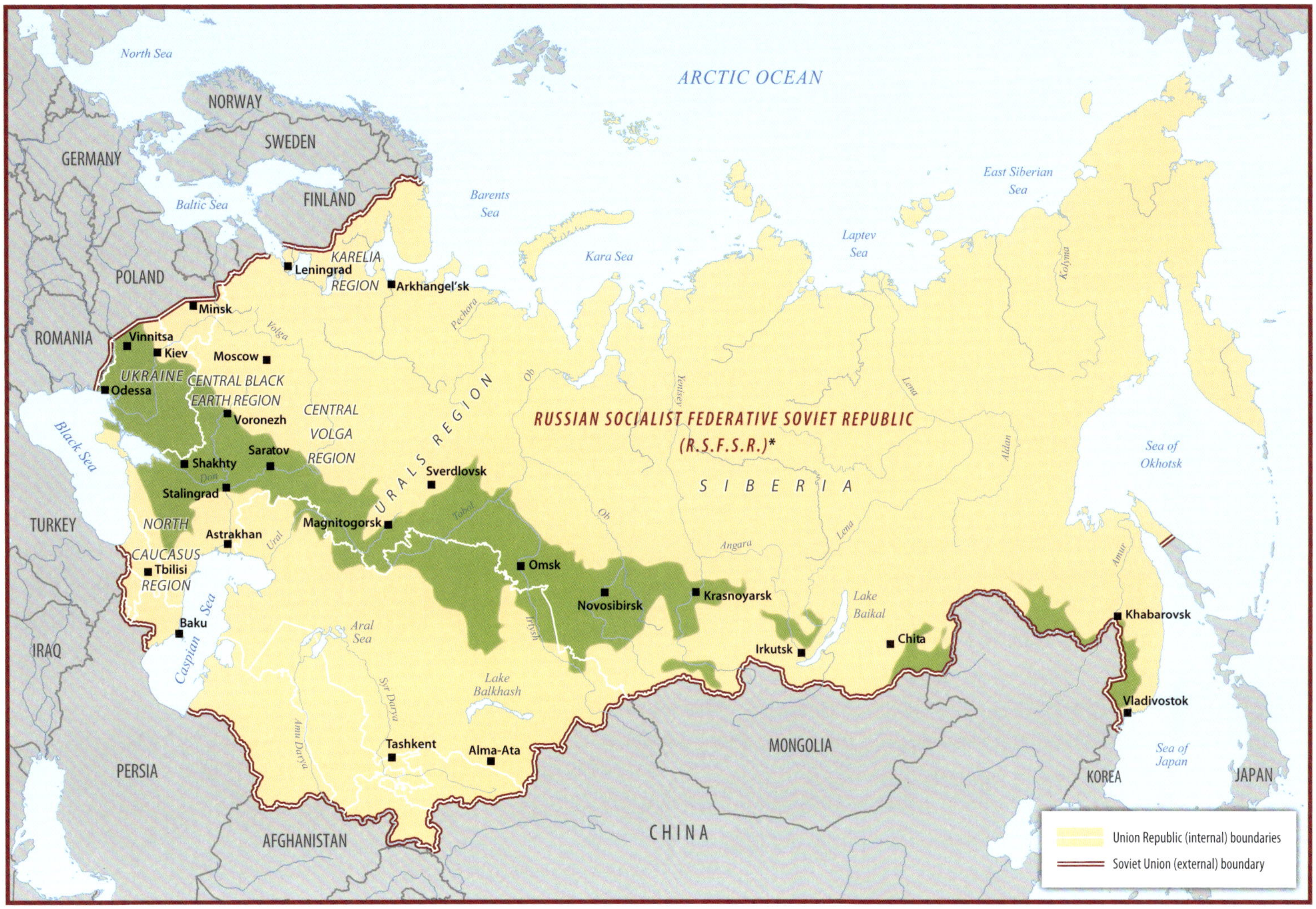

* Country name changed to Russian Soviet Federative Socialist Republic in 1937.

Map 1 Major grain-producing areas in the USSR

and officials professing their loyalty to the state, denying they had ever exploited labor, and insisting on their working-class credentials and ancestry. Many peasants sold their only cow or horse to pay their back taxes and avoid having all their property confiscated and auctioned off. Others divided up their landholdings and property among family members to avoid appearing well-off. Villagers spoke of a return to the cruelty of the civil war era.

The Bolshevik leaders had always been wary of what they called "bourgeois specialists." From the mid-1920s, many statisticians, economists, and demographers with an independent cast of mind had been quietly removed from positions of influence. On March 10, 1928, sixty engineers were arrested in the North Caucasus town of Shakhty and charged with sabotaging Soviet industry at the behest of Western capitalists. At a meeting of the Moscow Party organization on April 13, Stalin used the Shakhty affair to depict the Soviet Union as beset by terrible dangers both within and without. "We have internal enemies," he intoned. "We have external enemies. This, comrades, must not be forgotten for a single moment."[10] The "procurement crisis" was an attack launched by "capitalist elements" in the countryside. The Shakhty sabotage was an attack undertaken by "international capital and its agents in our country." Brought to trial the following month, thirty-four of the alleged saboteurs were sentenced to prison for between one and ten years. Five were executed. A key role in this episode was played by the secret police, whose budget, personnel, and responsibilities steadily expanded in these years. Many other specialists were purged from government agencies, undoubtedly as scapegoats to deflect attention from economic failures. Popular anxiety about internal and external enemies, including peasants allegedly withholding grain, sharpened.

So did the pressure on the peasantry, despite the fact that the party plenum in early April denounced coercive methods of grain collection. Stalin even joined in the chorus, though he insisted the abuses were merely deviations from the party line. Such was now his modus operandi: pretending to agree with the moderates but then implementing harsh policies as he saw fit.

A progressive tax on farm income, coupled with a wealth surtax, both adopted on April 21, aimed at economically ruining the "kulak." The London *Times* quoted Stalin admitting that three years before they had "needed the energetic richer peasant to extend the cultivated area." Now that he had completed this work, "we can safely destroy him, using the surplus land to form collective farms, cooperative farms, and big state farms."[11] The press frequently warned peasants to watch for the wiles of "kulaks" trying to worm their way into the roughly 23,000 collective farms that had been created by early May. "They did not give your collective farm a tractor? I have one and you can use it. I will join your farm."[12] In fact, better-off peasants traditionally helped their less fortunate brethren in all sorts of ways, such as lending them food and money in hard times. Village solidarity was very strong.

Tensions rose in the main grain-producing regions. In Ukraine, the North Caucasus, Siberia, the Urals, and along the Central Volga from April 15 to May 1, 1928, some 140 mass peasant protests broke out involving 32,500 people.[13] Many protesters were women, commencing a pattern that would become familiar over the next few years. The authorities throughout the country, for their part, arrested 6,794 people in April for "speculation" in grain and other commodities. In the Russian Republic alone, 5,597 people were tried, including 1,700 officially recorded as middle and poor peasants.[14] More and more

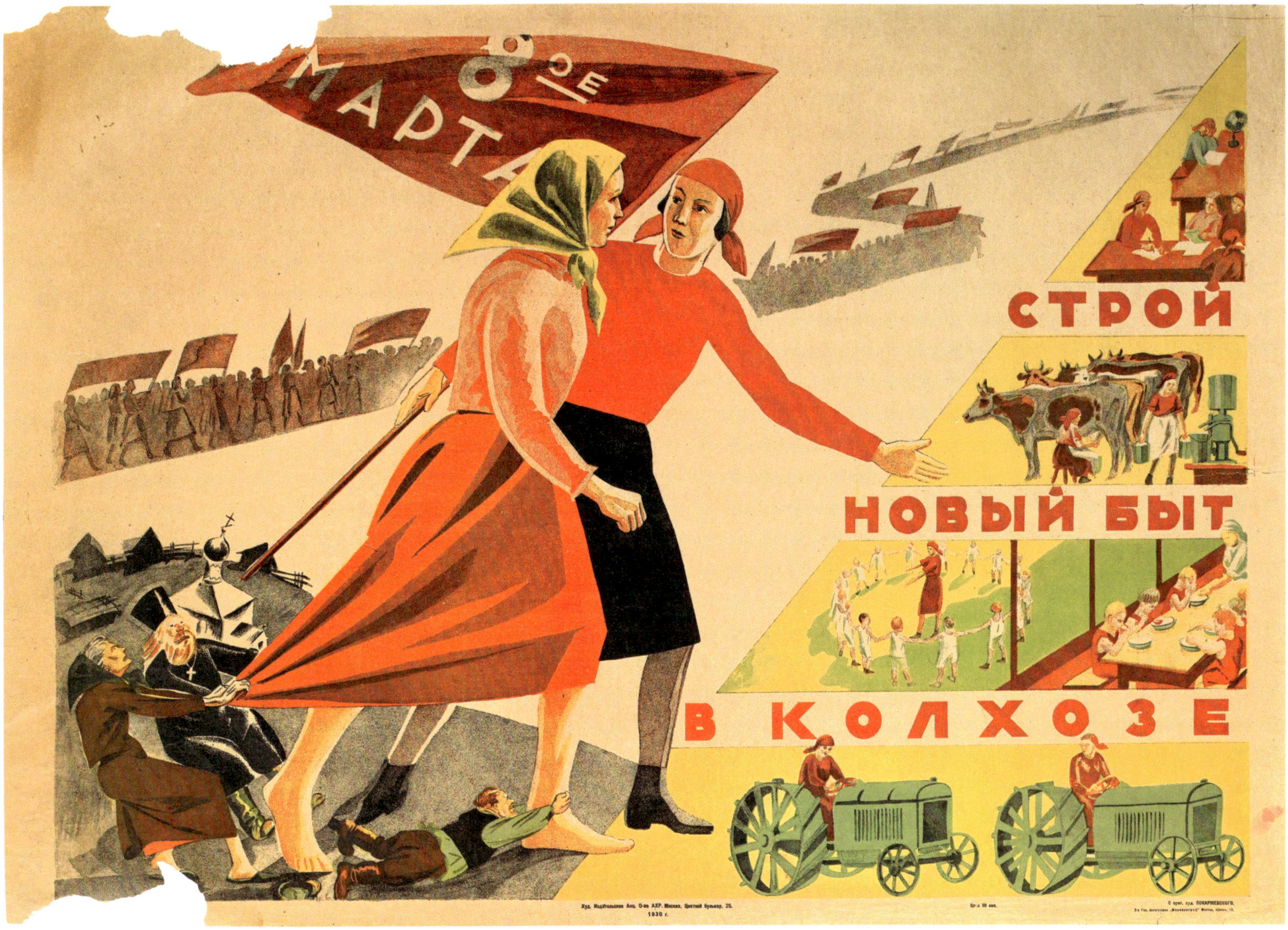

Fig. 9 P.D. Pokarzhevsky. "Build a new life in the collective farm." 1930. The banner reads, "March 8," which is International Women's Day.

arrests were made for counterrevolutionary activity, which article 58 defined as "anti-Soviet and counterrevolutionary propaganda and agitation" (paragraph 10).

The collections still failed to yield enough grain, and a further round of grain procurements took place in May and June 1928 with still more coercive methods. Grain collection in June was more than three times higher than the previous month, yet still the plan went unfulfilled. Reports of hunger came in. Relations with the countryside plunged to their lowest level since the civil war. Worse still, breadlines and unemployment riots revealed mass discontent in the towns as well. To stave off disaster, the party leadership took the humiliating step in June of importing grain—250,000 tons by August.[15]

The party plenum (July 4–11) was a jumble of contradictions. Bukharin and the moderates vigorously advocated a set of policy shifts designed to ease pressure on the peasantry, including an end to arbitrary confiscations and price increases for grain procurements. A master tactician,

Stalin approved almost all their proposals. Of course, his agents in the field would continue to apply harsh methods. The moderates apparently doubted his sincerity, but having made a fetish of party unity, they could not easily denounce him in the open. Well might they have doubted.

On the second day of the plenum, Stalin made his notorious reference to a "tribute" from the peasantry. To industrialize, he argued, big investments were needed. The USSR was an anti-imperialist country and therefore "cannot and must not engage in colonial robbery, or the plundering of other countries in general." Nor could it agree to "enslaving loans from abroad." Yet how to pay for grandiose projects like the Turkestan Railway and the Dnieper Hydro-Electric Power Station—each costing hundreds of millions of rubles? Only one source of investment was left: "internal accumulations." The working class, he went on, "creates values and advances our industry." Apparently no more was expected. The peasantry, for its part, "not only pays the state the usual taxes, direct and indirect; it also overpays in the relatively high prices for manufactured goods . . . and it is more or less underpaid in the prices for agricultural produce." Everyone at the plenum knew he was right: even without special coercive methods, the peasants were being squeezed on all sides. But now Stalin went for the jugular:

> This is an additional tax levied on the peasantry for the sake of promoting industry, which caters for the whole country, the peasantry included. It is something in the nature of a "tribute," of a supertax, which we are compelled to levy for the time being in order to preserve and accelerate our present rate of industrial development, in order to ensure an industry for the whole country, in order to raise further the standard of life of the rural population and then to abolish altogether this additional tax, these "scissors" between town and country.
>
> It is an unpalatable business, there is no denying. But we should not be Bolsheviks if we slurred over it and closed our eyes to the fact that, unfortunately, our industry and our country cannot at present dispense with this additional tax on the peasantry.[16]

Bukharin and the moderates were outraged by such talk, because the term *tribute* evoked images of Genghis Khan and of the harshest rule of the Muscovite tsars. But surely Stalin was merely calling a spade a spade, for, whatever the rhetoric about *smychka*, the party-state was indeed squeezing the peasantry, exacting a "tribute" from it. What else can one call buying produce well below market prices and selling (scarce) manufactured goods well above? He went on to argue that emergency measures deployed in grain collection had "saved the country from a crisis of our whole economy," but he hastened to add that they were "emergency" measures needed only in a time of crisis. What Stalin—and the others—did not realize, however, was that the premises on which they were building socialism were producing one crisis after another: it is simply impossible to pay producers less than what the law of supply and demand requires and expect them to continue producing and delivering.

Rural unrest abated for the rest of the year, as did coercive measures, but grain collections were down—40 percent less in July and August than in the same months a year before. And pressure was kept up on the peasants. Beginning on September 14, 1928, they had to pay a "milling levy" (*garntsevyi sbor*). Nor did the price paid for their grain increase much. Other taxes went up on most households as well. The first Five-Year Plan, which officially commenced on October 1, foresaw pouring 78 percent of all capital investment into industry. Precious little was left over for agricultural machinery, chemical fertilizer, and other assistance to agriculture.

Two factors now came to Stalin's assistance. On the penultimate day of the July plenum, Bukharin paid a secret visit to a disgraced Kamenev, to confer about joining forces against Stalin, who naturally used information about this meeting to his advantage in the impending battle within the Politburo. Second, a police report for June and July 1928 indicated that the less prosperous peasants backed the procurement campaign, presumably in large part because official policy since January had involved distributing 25 percent of confiscated grain to the poor.[17] At the local level, many officials had begun to realize that social pressure—of the have-nots against the haves—might be the best way to squeeze grain from the countryside.

In any event, paltry collections—by late December only 300 million poods of grain, as against an anticipated 500[18]—compelled the authorities to introduce bread rationing in the towns, probably the first such instance in any country in peacetime.

Stalin now argued for a return to harsh methods, but this time built around a "social-pressure" or "social-incitement" approach, which became known as the "Ural-Siberian method." He and his supporters carefully laid the groundwork. On January 3, 1929, the Politburo ordered the Commissariat of Justice to speed up the "repression of kulak terrorists."[19] On January 6, an order went out to local Soviets threatening dire consequences for failure to achieve larger collections. For the first time, grain collection was called a "front." Military police went through the ranks, weeding out by mid-January 1929 "socially dangerous" and "class-alien" soldiers, many of them sons of kulaks.[20] All the previous year, families had written to their conscripted sons complaining of coercive grain collection and poor living conditions. The authorities obviously wanted to protect the military from such fonts of disgruntlement. On February 2, the political police forbade deploying troops in anti-"kulak" operations and ordered the interception and analysis of all letters sent to and from soldiers. The party leaders were clearly worried about disloyalty in the ranks.

On January 17, a Politburo order went out—and was much imitated by local officials—to press rural assemblies to vote for heavier taxes on the better-off. Sometimes taxes were doubled. Local officials were encouraged to organize boycotts of those who fell behind in their payments. Those subject to boycott could face public denunciation as enemies of Soviet power, exclusion from cooperatives, shunning, insults, denial of access to manufactured goods and public services like medical and veterinary care, and receipt of threatening notes. One such note in Siberia read, "Most malicious enemy of Soviet power so and so, we know you have grain. . . . Don't wait for prices to go up, give up your surplus to the state."[21] In late January 1929, R. I. Eikhe (1890–1940), the chairman of the Siberian Executive Committee, urged rural activists to "hound the kulaks for their withholding of grain surpluses, organize against them a social cooperative boycott."[22] The percentage of confiscated grain promised to poorer villagers presumably inspired enthusiasm. The Siberian authorities soon issued rules for conducting boycotts: poor and middle peasants who had yielded up their surplus should target three to five families in the village to pressure them to do likewise.

Despite such efforts, countrywide grain collections declined in January by 36 percent over the previous month and in February by 20 percent, in part because of a widening gap between state and market prices. The

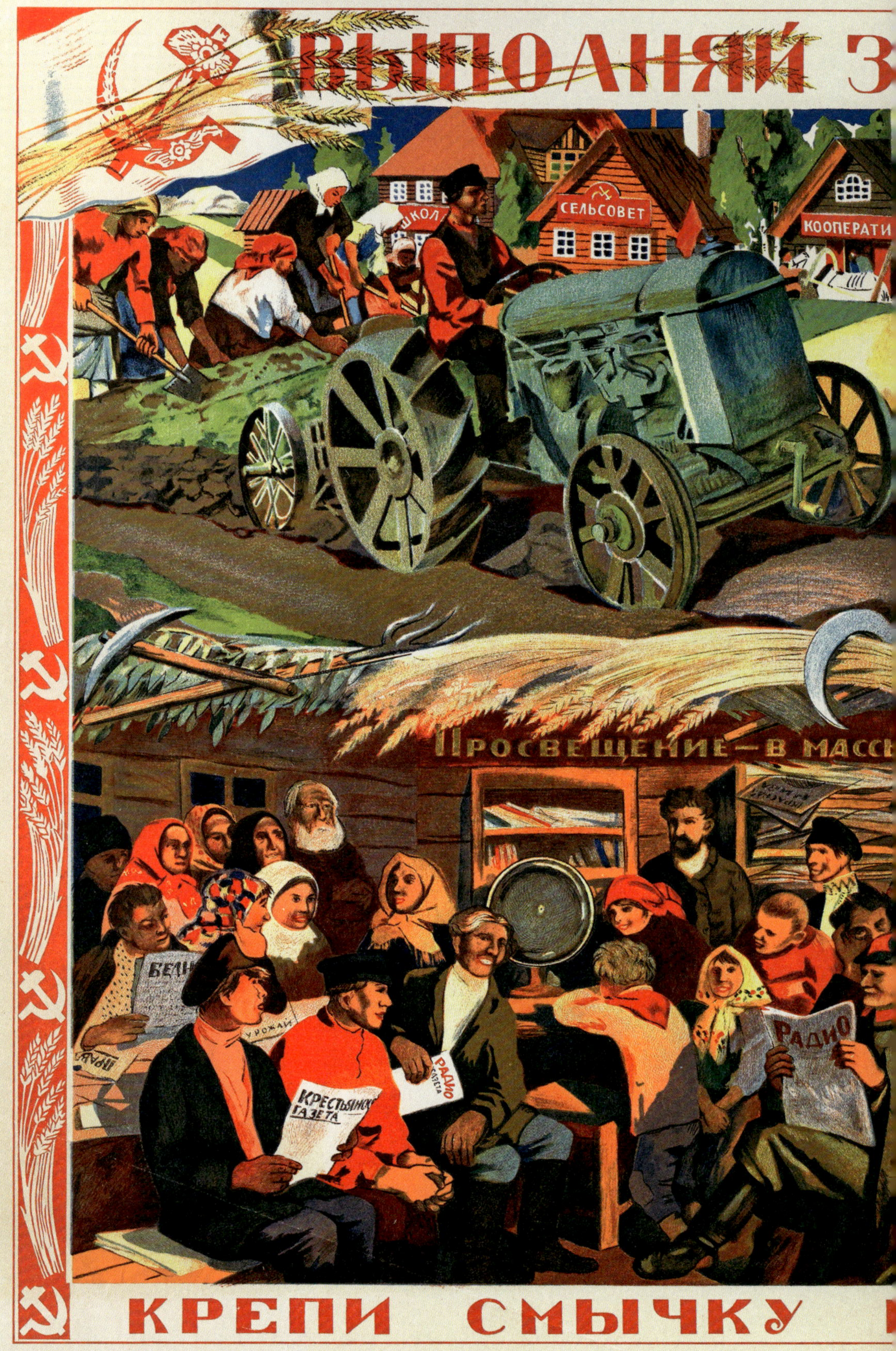

Гиз № 23459/л. Ленинградский Гублит № 47597. Тираж 35.000. ГОСУДАРС

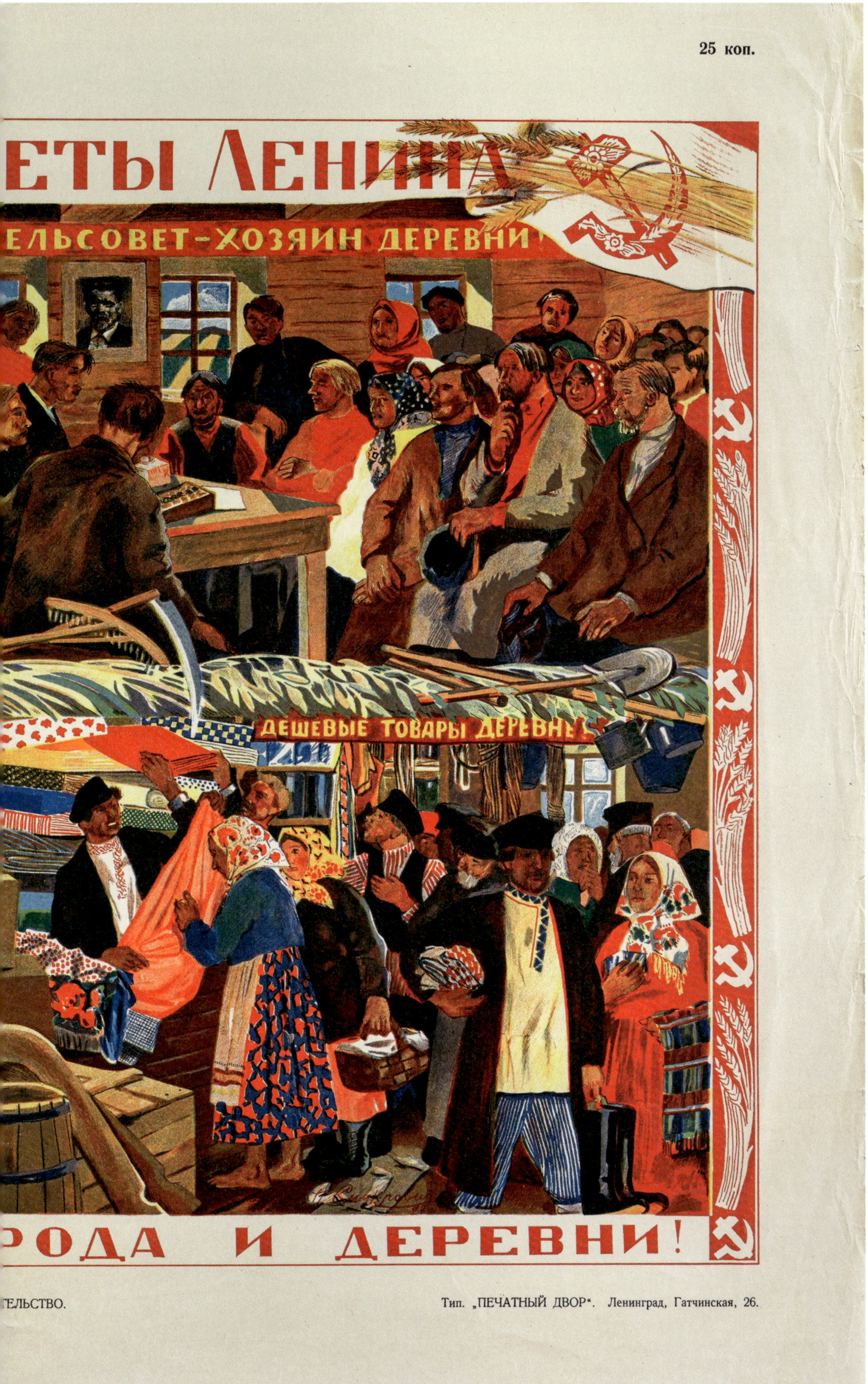

Fig. 10 P.I. Smukrovich. "Fulfill Lenin's behests. Strengthen the *smychka* of town and country!" 1929.

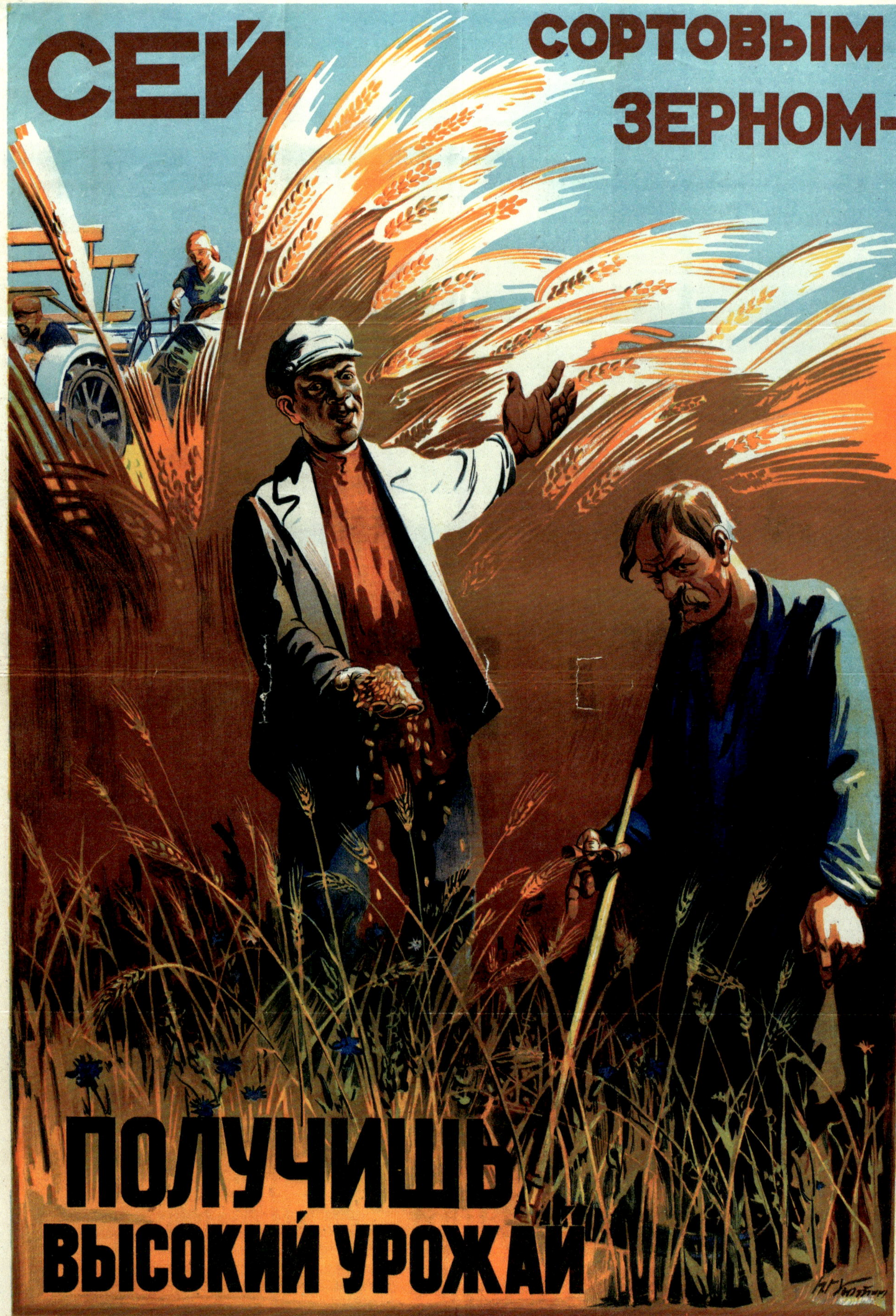

20 коп.

Главлит №А–71145. Р. 30. № 106. Зак. 1287. Гиз 40922. Тир. 40000. Из них 7500 на украинском языке.

ГОСУДАРСТВЕННОЕ ИЗДАТЕЛЬСТВО
МОСКВА 1930 ЛЕНИНГРАД

Типо-литография Госиздата «Красный пролетарий» Москва, Пролетарская 16.

government printed more money, further eroding purchasing power. Strikes and absenteeism in factories grew frequent.

Grain collection seemed to turn a corner in March 1929 in the Ural region and in Siberia. Village boycotts and shaming rituals directed against the well-to-do were ramped up, as was economic pressure. In particular, failure to pay imposed assessments could result in their fivefold increase, a procedure that apparently emerged at the local level but received Moscow's sanction on April 5 and again in late June. Not seldom, local party activists and poor peasants excitedly despoiled "the rich" down to the shirts on their backs. The takings were often roughly two-thirds from the better-off and one-third from "middle peasants." Positive results made the career of the Siberian party leader, S. I. Syrtsov (1893–1937), who became a candidate Politburo member in May. In mid-March, Lazar Kaganovich (1893–1991), one of Stalin's right-hand men, arrived in Sverdlovsk in the Urals and approved their version of the experiment. On March 20, the Politburo approved the Ural region initiative and proposed to apply it in Kazakhstan, Siberia, and elsewhere. Stalin was most pleased. On March 11, he wrote to the party boss of the grain-producing North Caucasus region, Andrei Andreev (1895–1971), another of his henchmen, that every extra million poods "is precious to us."[23] At the party plenum, which met on April 16–23, Stalin referred to this set of procedures as "the Ural-Siberian method of grain procurement." In those regions, grain collections in April were up 273 percent over the same month in 1928.[24]

Stalin was slowly triumphing politically as well. The plenum condemned the "right deviation" (though Bukharin was revealed as its leader only in August). The Sixteenth Party Conference on April 23–29 approved increased Five-Year Plan targets, including the full collectivization of all grain-producing regions by early 1932 at the latest—targets the rightists and nearly all experts considered unfeasible. The conference also celebrated an alleged movement of "middle peasants" into collective farms and sharply denounced the "kulaks."

Yet the party leaders were worried. Secret police reports in May 1929 pointed to increased "kulak" and religious "counterrevolution" in the countryside. On May 21, the Council of People's Commissars carefully spelled out such characteristics of a "kulak" household as hiring labor, possessing a mill or mechanical farm implements, renting out large buildings, trading, or lending money. Yet even many poor peasants did not seem to view such activities as detestable. In fact, they often mistrusted the state but looked up to the "kulaks." As Mikoian himself admitted, speaking at the June 1929 Seventh All-Union Collective Farm Council, "When questions arise about planting and plowing, the middle peasant does not go to the poor peasant for an answer because the poor peasant's farm is worse than his own. He is more likely to look to a farm that is well supplied, where the horses are handsome and the harvest is good."[25] The very success of the "kulaks" posed a serious threat to the Soviet regime, which had yet to prove it could outproduce them.

Aside from hostility to grain collection, most rural folk surely disapproved of increasingly harsh antireligious policies. The decree on religious associations of April 8 criminalized most forms of mutual aid, association, education, and cultural activity for religious purposes. Since these strictures contradicted the right to "religious and anti-religious propaganda" enshrined in the Constitution of 1924, an amendment was adopted on May 22, 1929. The constitution now declared that

Fig. 11 N. G. Kotin. "Sow high-yield seeds and obtain a big harvest." 1930.

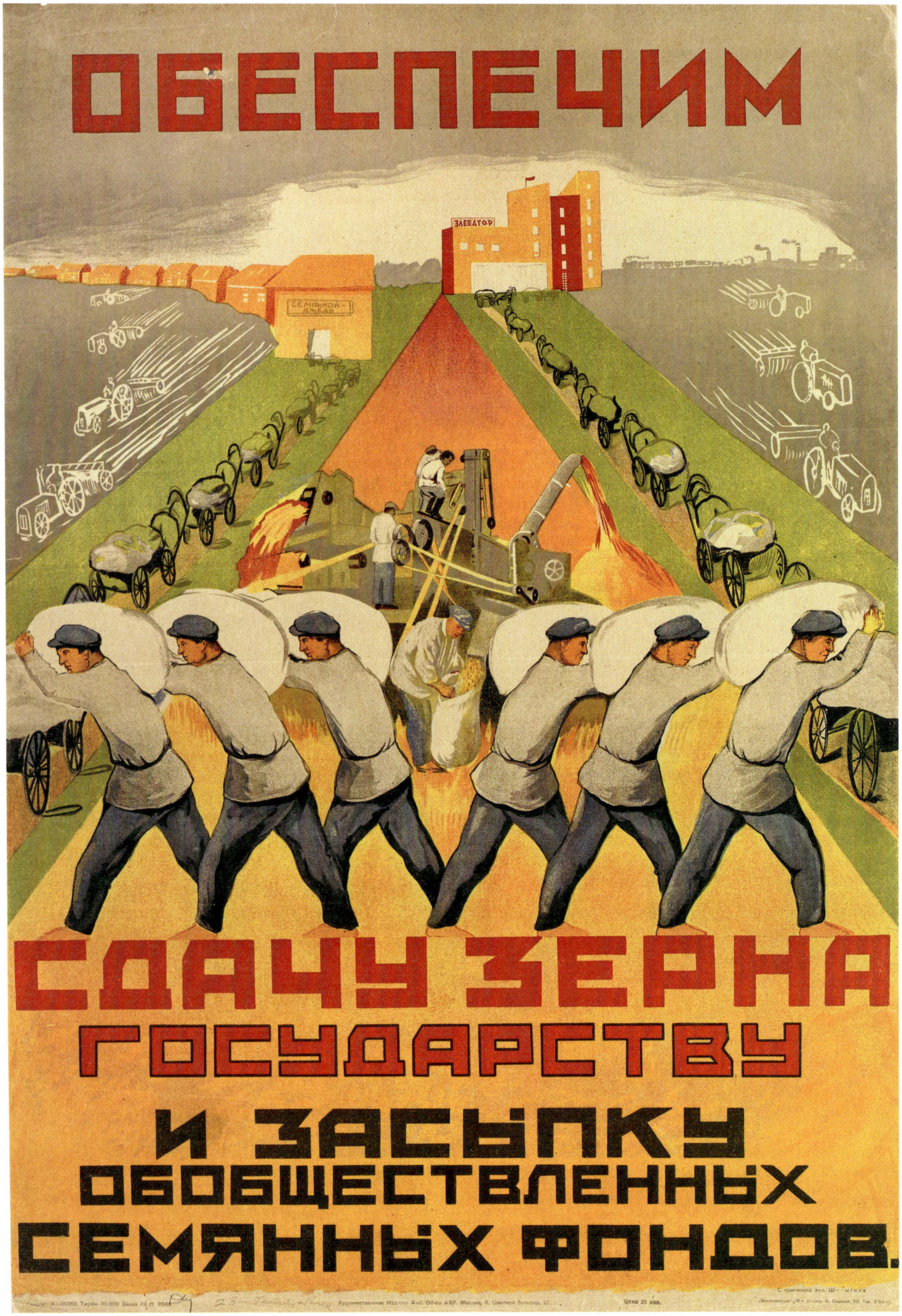

◀ **Fig. 12** F. S. Shurygin. "We will ensure the delivery of grain to the state and to socialized seed funds." 1920–28?

▶ **Fig. 13** Artist not identified. "Let us create indivisible funds for the acquisition of tractors." 1930. "Gather horsehair, livestock horns, human hair, rags, and bones, for which abroad they exchange tractors and combines."

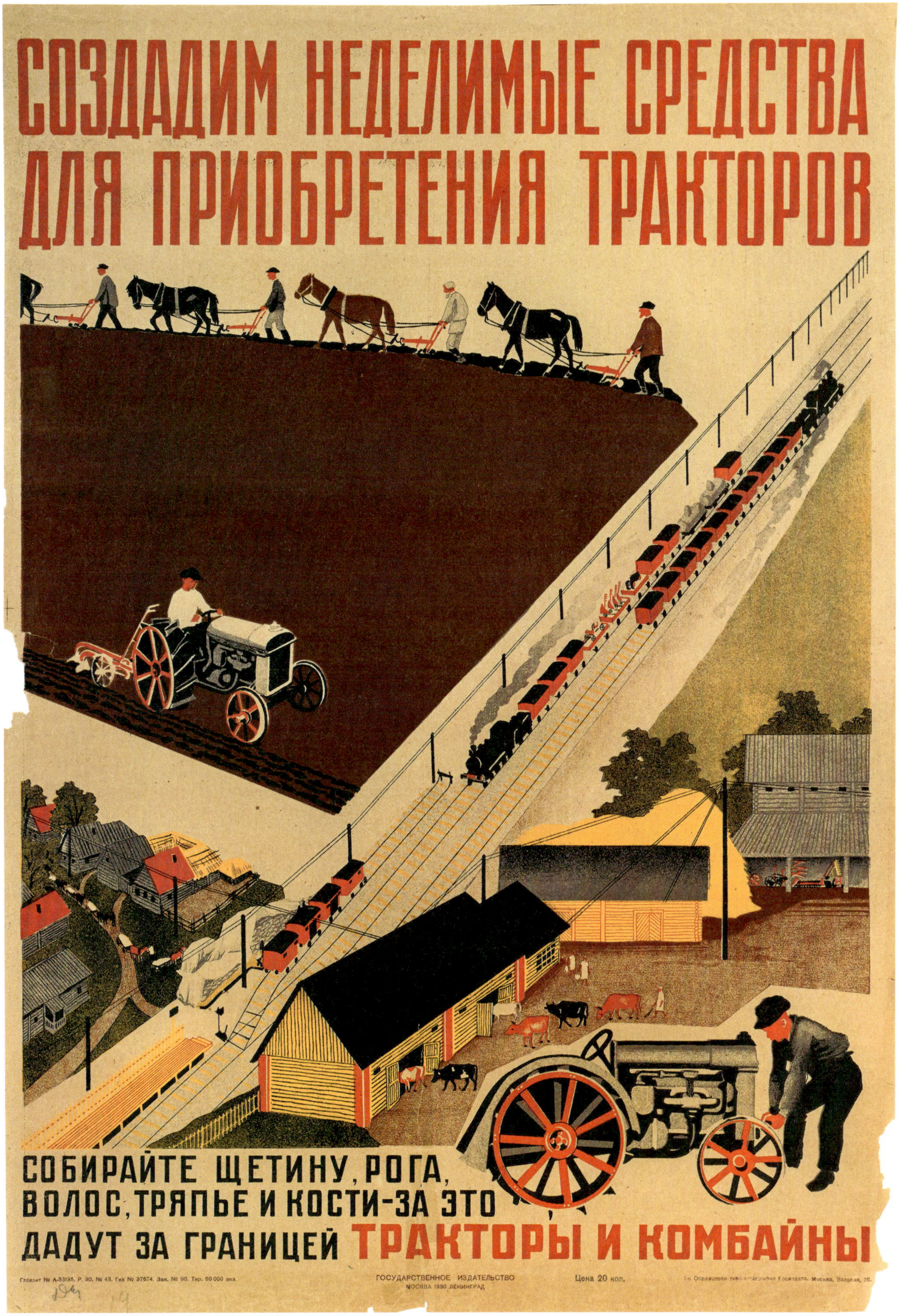
СОЗДАДИМ НЕДЕЛИМЫЕ СРЕДСТВА
ДЛЯ ПРИОБРЕТЕНИЯ ТРАКТОРОВ
СОБИРАЙТЕ ЩЕТИНУ, РОГА,
ВОЛОС, ТРЯПЬЕ И КОСТИ-ЗА ЭТО
ДАДУТ ЗА ГРАНИЦЕЙ ТРАКТОРЫ И КОМБАЙНЫ
Главлит № А-53135. Р. 30. № 43. Гиз № 37674. Зак. № 90. Тир. 60 000 экз.
ГОСУДАРСТВЕННОЕ ИЗДАТЕЛЬСТВО
МОСКВА 1930 ЛЕНИНГРАД
Цена 20 коп.

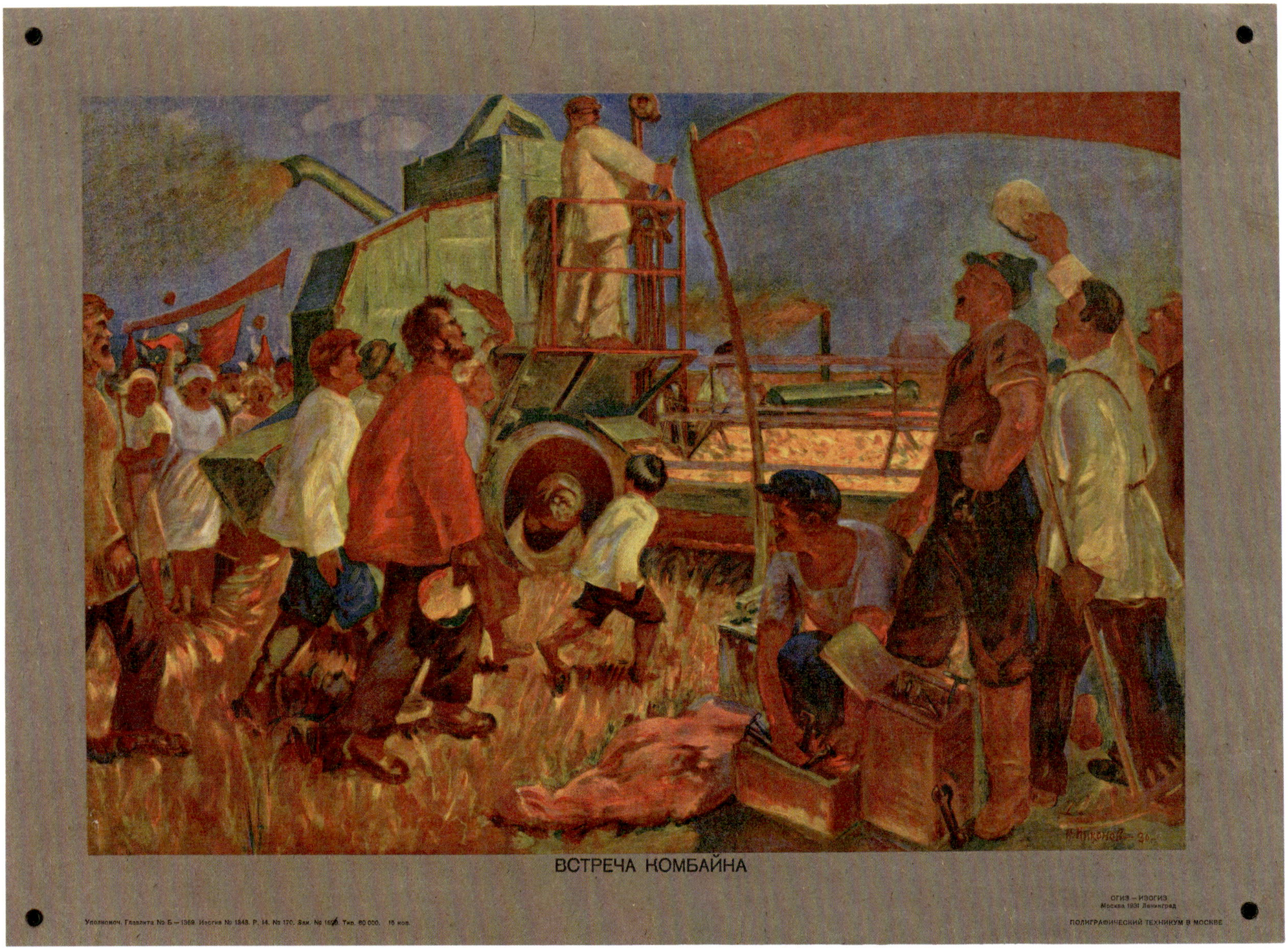

Fig. 14 N.M. Nikonov. "Encounter with a combine." 1931.

"no activity beyond the limits of the mere satisfaction of one's religious aspirations is permitted."[26] On August 26, the government abolished Sunday as a day of rest. The Central Council of Trade Unions heralded the move as facilitating "the eradication of religion."[27] As unrest and rumors broke out across the villages, antigovernment rhetoric gradually took on more and more apocalyptic overtones. For many, the Antichrist and Armageddon were coming.

The Antichrist would drive a tractor. For months, Stalin had been touting the dramatically greater efficiency he expected to be achieved by larger farms. Such farms would be highly mechanized, with tractors systematically replacing horses. Thus on June 5 the government gave instructions to create a network of machine tractor stations (MTS) and that month reportedly put in an order to buy 6,750 tractors in America.[28] The Russians, one can argue, had a fetishistic or iconic attitude toward technology—as if simply building and deploying machines throughout society could radically transform reality. The Soviet agrarian economist and utopian thinker, Alexander Chaianov (1888–1937), enthused in 1929 that "the triad of American machine techniques—the *crawler*, *combine*, and *truck*—brings into the present techniques of agriculture, so far as we can judge from the available material, and in the organization of grain production, a technical revolution which in importance must be compared with the appearance of the steam engine in industry."[29] The Bolsheviks shared his excitement. For many of them, only mechanization and economies of scale could offer a compelling alternative to "kulak" success.

Yet the tractors and combines were slow in coming, repeatedly broke down when they arrived, required organizational and technical expertise that was habitually lacking, and did not fit easily with prevailing Soviet culture, as the many hundreds of American experts invited to help mechanize Soviet agriculture often found.

Repression and coercive methods served as a handy supplement. Decrees of June 27 and 28, 1929, spelled out disabilities affecting "kulaks" and gave collection agents a variety of new tools. Article 61 of the criminal code was amended to formally authorize imposing fivefold tax obligations even on first-time offenders. Of course, few households could afford such a gigantic fine, when in most cases they had been unable to pay the original sum. For a second offense, one could face up to two years in prison, confiscation of property, and even deportation. This was a big step toward "dekulakization." "Kulaks" and "non-laboring elements" also lost their voting rights, which kept them out of soviets, cooperatives, and other civic organizations. Throughout the summer, perhaps two hundred thousand party activists, Komsomol members, trade-union organizers, and factory workers were sent to the countryside to assist with grain collection. On August 7, *Pravda* declared the USSR to be "on the road to mass collectivization."[30]

On August 15, 1929, following a recommendation by Stalin, the Politburo ordered the secret police to root out urban grain "speculators" and rural traders with connections to them. A directive of September 9 went farther. Peasants were forbidden to sell grain on the free market, even after they had met their delivery quota to the state. A new round of arrests in connection with grain procurement snared 1,187 people as of September 7, and 2,465 eleven days later (and 28,344 by November 4), according to incomplete data.[31] Hundreds of indictments were brought against officials for inaction and abetting speculators. From September 20, major newspapers demanded measures to "break the resistance of the kulak"[32] and to

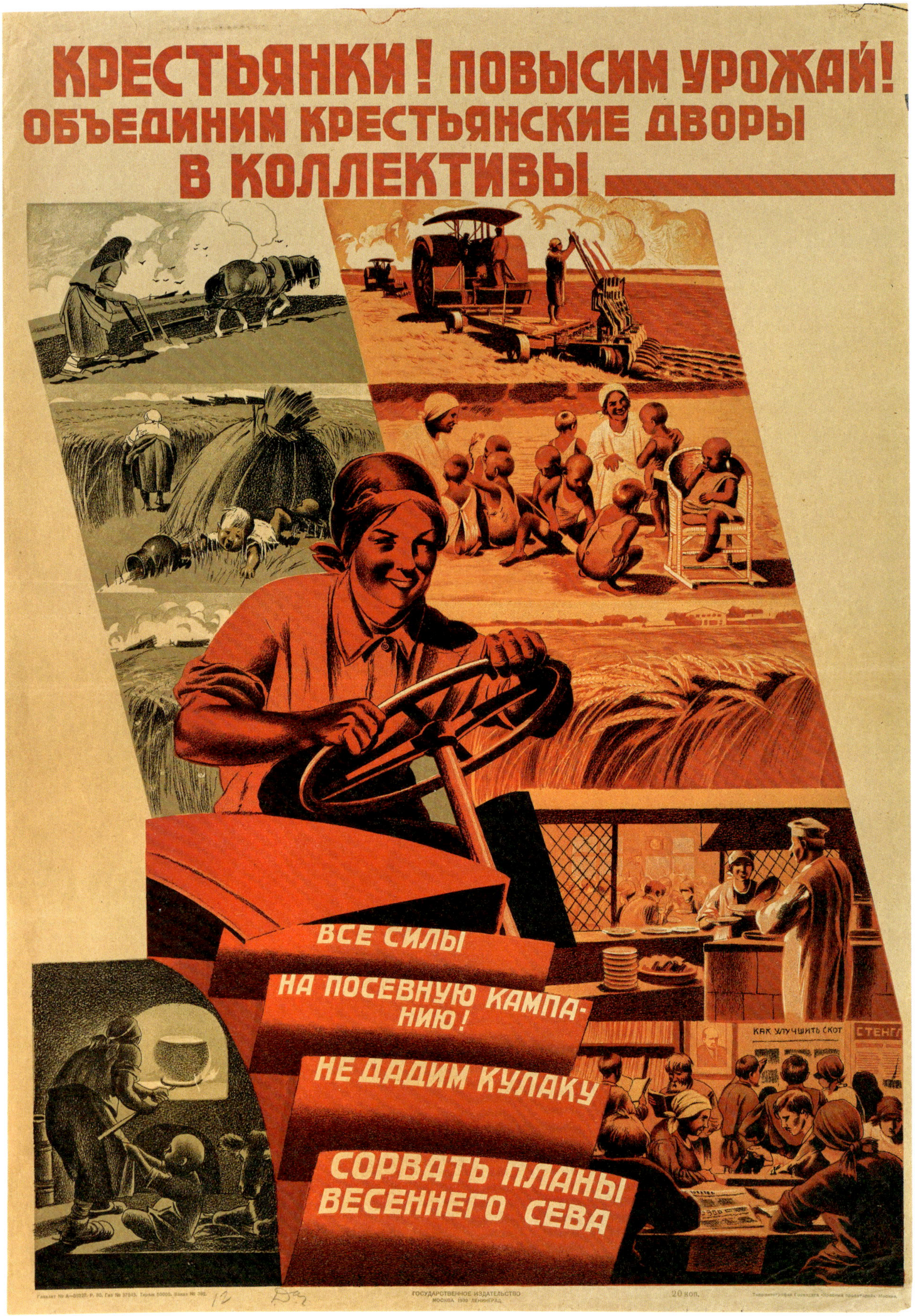

◀ **Fig. 15** Artist not identified. "Peasant women! Let's increase the harvest! Let's join all peasant households into collectives." 1930. "All our strength into the sowing campaign! We won't let the kulak wreck the spring sowing plan."

▶ **Fig. 16** L.T. Chupiatov. "Let's organize a spirited sowing campaign." 1930.

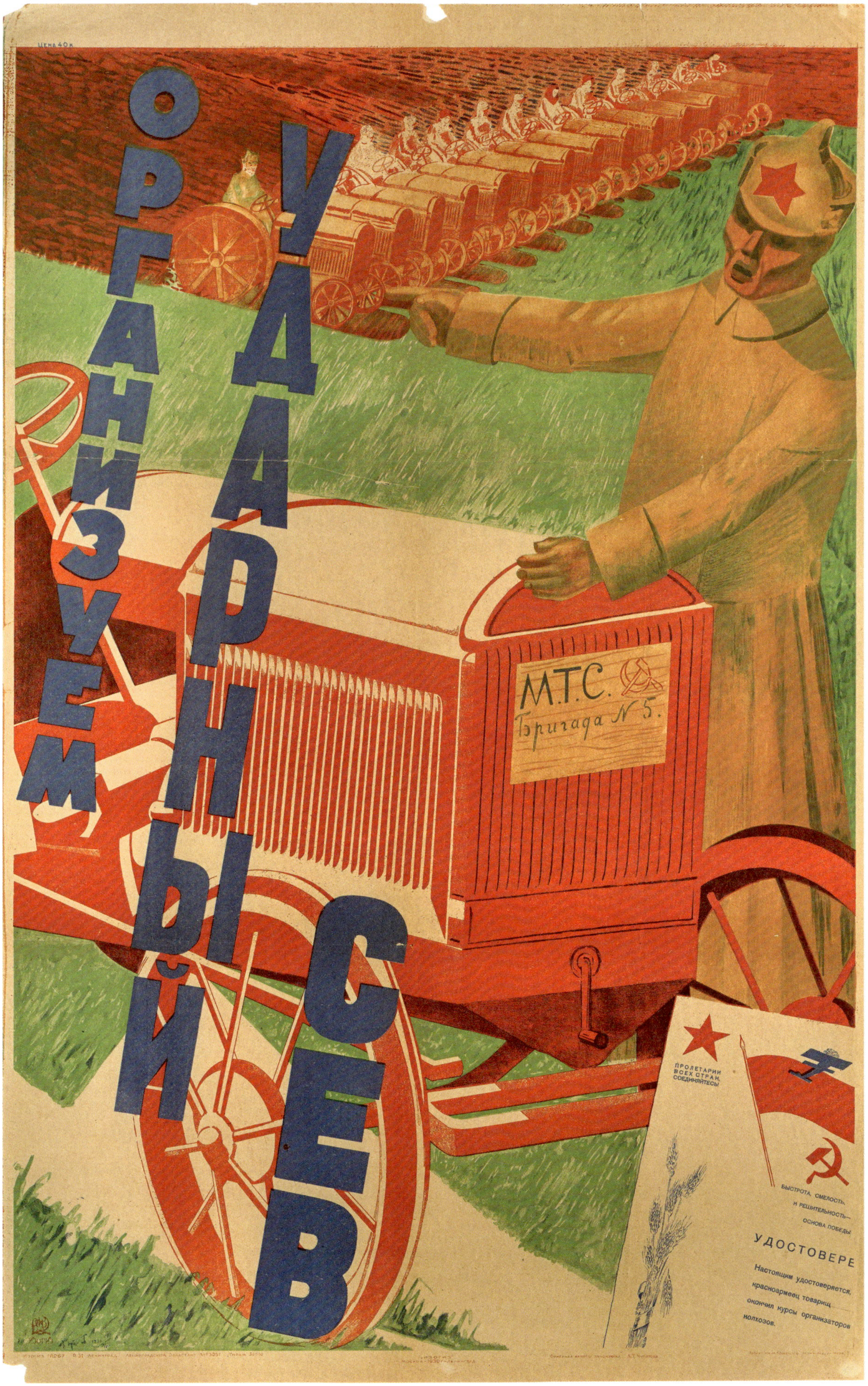

Цена 40 к.
ОРГАНИЗУЕМ
УДАРНЫЙ
СЕВ
М.Т.С.
Бригада № 5.
ПРОЛЕТАРИИ ВСЕХ СТРАН, СОЕДИНЯЙТЕСЬ!
Быстрота, смелость и решительность— основа победы
УДОСТОВЕРЕ
Настоящим удостоверяется, красноармеец товарищ
окончил курсы организаторов колхозов.

socialize peasant livestock. Revolutionary zeal was taking hold, as party activists called for completing the first Five-Year Plan in four years. Unprecedented success in grain procurements breathed fire into already optimistic expectations.

The next stage in getting control of the countryside and the grain supply was pushing more peasants into collective farms. There was gradual movement in this direction. According to official figures, the number of households in collective farms rose from 1 million (or 3.9 percent) in June 1925 to 2.19 million (or 7.5 percent) in September[33]—a fast tempo, relatively speaking, but a puny absolute number. It turned out, moreover, that most who joined owned neither farm implements nor livestock or had sold them prior to joining. Better-off peasants were in fact selling and slaughtering livestock right and left. Why not? A decree of November 14 empowered officials to auction off rural property for tax delinquency. That month, the Middle Volga regional party boss, Mendel' Khataevich (1893–1937), warned that total livestock by spring would drop 20–25 percent. (In reality, herds had already shrunk much more in some regions.) Peasant rebellions and attacks on officials proliferated, too. Peasants in grain-producing regions also began to plant less—at least 10 percent less. Again, why plant if one's crops will be taken? Discontent in the military grew stronger as soldiers heard horror stories from back home. Nor could most of them understand the need to do away with "kulaks" entirely. "The kulaks do not need to be annihilated," as one put it; "they need to be taken into the collective farm so they can show the poor people how they ought to work."[34] Not surprisingly, when forced collectivization began in earnest in 1930, the military was little involved.

On November 3, 1929, on the eve of the twelfth anniversary of the Bolshevik Revolution, Stalin gave a crucial speech entitled "A Year of Great Change," which *Pravda* printed four days later. He argued that it was time to halt the retreat commenced with the NEP under Lenin's leadership, thanks to "unprecedented success in agricultural development" in 1929. Most important, "the peasants are joining the collective farms; they are joining by whole villages, volosts, districts. . . . the *middle peasant is joining the collective farm*."[35] Of course, such utterances were absolutely untrue. But who would dare challenge them? The Central Committee plenum, which met on November 10–17, disgraced Bukharin and expelled him from the Politburo. True, Syrtsov warned that the collectivization efforts so far had involved "a divergence between quantitative growth and qualitative organization of large production. If we do not immediately take measures to strengthen the collective farms, the entire enterprise could break down." Stalin snapped back: "So you think everything can be 'organized in advance'?"[36] The plenum went ahead and voted to speed up the pace of collectivization. Stalin supporters spoke publicly and privately of something like a military campaign against the "kulaks."

A week after Stalin's fairly grandiose fiftieth birthday celebration on December 21 (the start of his cult of personality), he spoke at the All-Union Conference of Marxist Students of Agrarian Questions, announcing universal collectivization and "liquidation of kulaks as a class." In fact, he claimed, this was happening spontaneously on the initiative of ordinary peasants: "The collective-farm movement, which has assumed the character of a mighty and growing anti-kulak avalanche, is sweeping the resistance of the kulak from its path, is shattering the kulak class and paving the

Fig. 17 V. S. Korableva. "Come, comrade, join us in the collective farm!" 1930.

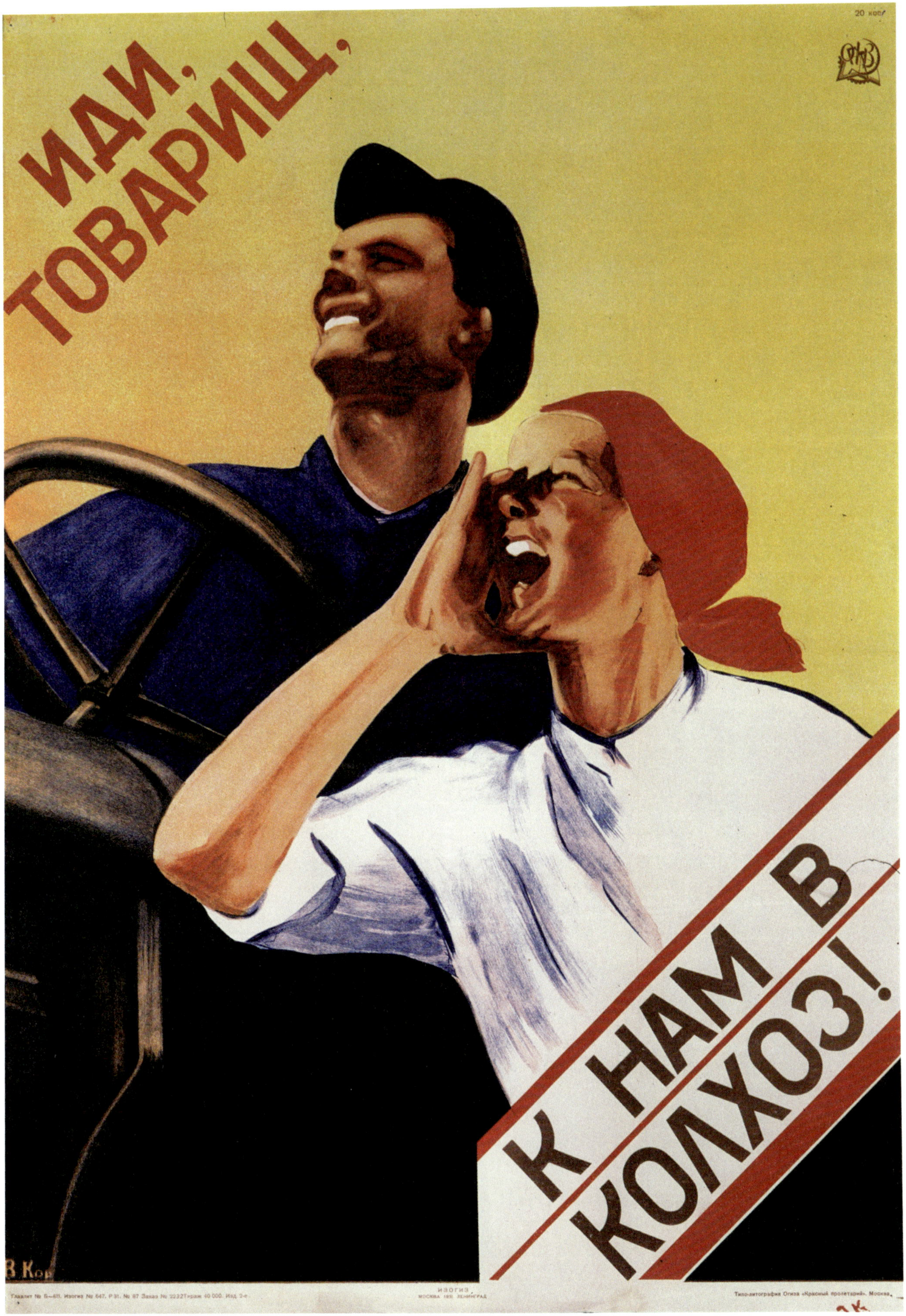
20 коп.
ИДИ, ТОВАРИЩ,
К НАМ В КОЛХОЗ!
В Кор
ИЗОГИЗ
Типо-литография Огиза «Красный пролетарий». Москва.

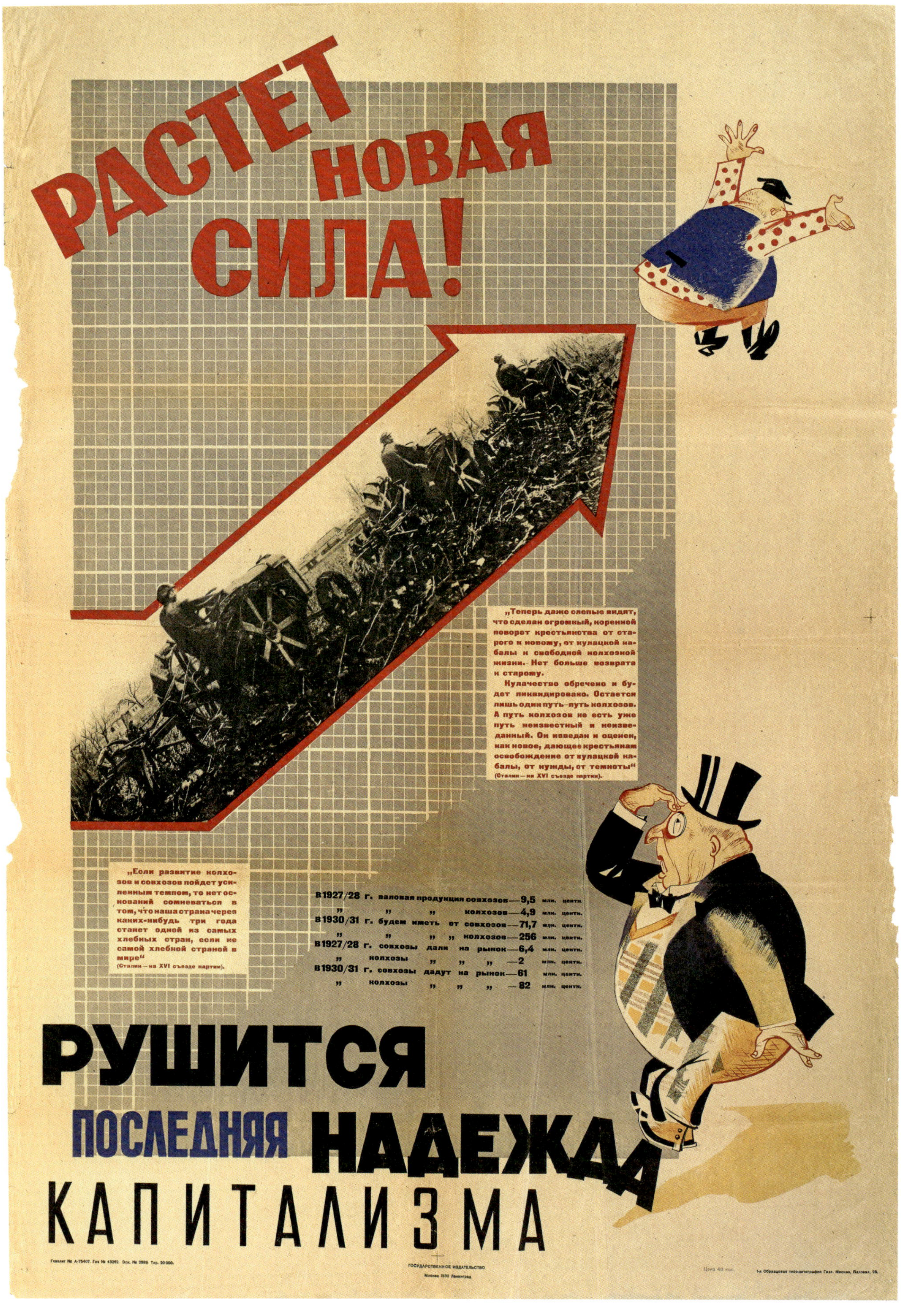
РАСТЕТ НОВАЯ СИЛА!
„Теперь даже слепые видят, что сделан огромный, коренной поворот крестьянства от старого к новому, от кулацкой кабалы к свободной колхозной жизни. Нет больше возврата к старому.
Кулачество обречено и будет ликвидировано. Остается лишь один путь—путь колхозов. А путь колхозов не есть уже путь неизвестный и неизведанный. Он изведан и оценен, как новое, дающее крестьянам освобождение от кулацкой кабалы, от нужды, от темноты“
(Сталин—на XVI съезде партии).
„Если развитие колхозов и совхозов пойдет усиленным темпом, то нет оснований сомневаться в том, что наша страна через каких-нибудь три года станет одной из самых хлебиых стран, если не самой хлебной страной в мире“
(Сталин—на XVI съезде партии).
в 1927/28 г. валовая продукция совхозов—9,5 млн. центн.
„ „ „ колхозов—4,9 млн. центн.
в 1930/31 г. будем иметь от совхозов—71,7 млн. центн.
„ „ „ „ колхозов—256 млн. центн.
в 1927/28 г. совхозы дали на рынок—6,4 млн. центн.
„ колхозы „ „ „ —2 млн. центн.
в 1930/31 г. совхозы дадут на рынок—61 млн. центн.
„ колхозы „ „ „ —82 млн. центн.
РУШИТСЯ ПОСЛЕДНЯЯ НАДЕЖДА КАПИТАЛИЗМА
Главлит № А-75407. Гиз № 43262. Зак. № 2589. Тир. 20 000.
ГОСУДАРСТВЕННОЕ ИЗДАТЕЛЬСТВО
Москва 1930 Ленинград
Цена 40 коп.
1-я Образцовая типо-литография Гиза. Москва, Валовая, 28.

Fig. 18 Artist not identified. "A new power is rising! The last hope of capitalism is crashing." 1930.

way for extensive socialist construction in the countryside."[37] He also asserted decisively that "kulaks," as "sworn enemies of the collective-farm movement," could never be allowed to join collective farms. The fact that the annual grain procurement plan had been overfulfilled by 9.2 percent doubtless buoyed the spirits and stiffened the resolve of Stalin and his closest comrades that they could storm any fortress. Conquering the peasantry meant everything to Stalin. As he wrote to Molotov, "If we can beat this grain thing, then we will prevail in everything, both in domestic and foreign policies."[38] To reports that famine was beginning to stalk the countryside as a result of excessive grain collections they paid little heed.

The rank-and-file and lower-level leadership of the Communist Party had had their "baptism of fire" during the civil war. They saw it as a glorious time of triumph over terrible odds and vile enemies. Capitalism was the worst enemy. Yet during the NEP, enriched "bourgeois" were showing off fur coats, diamond rings, fancy cars, and tables buckling under the weight of caviar and champagne. For what, then, had the revolution and the civil war been fought? When Stalin and his supporters decided to attack the nouveau riche, to overthrow the moderates in the Communist Party, to assault the church and religious believers, and to extract a real tribute from the peasantry, they would enjoy widespread support throughout the party.

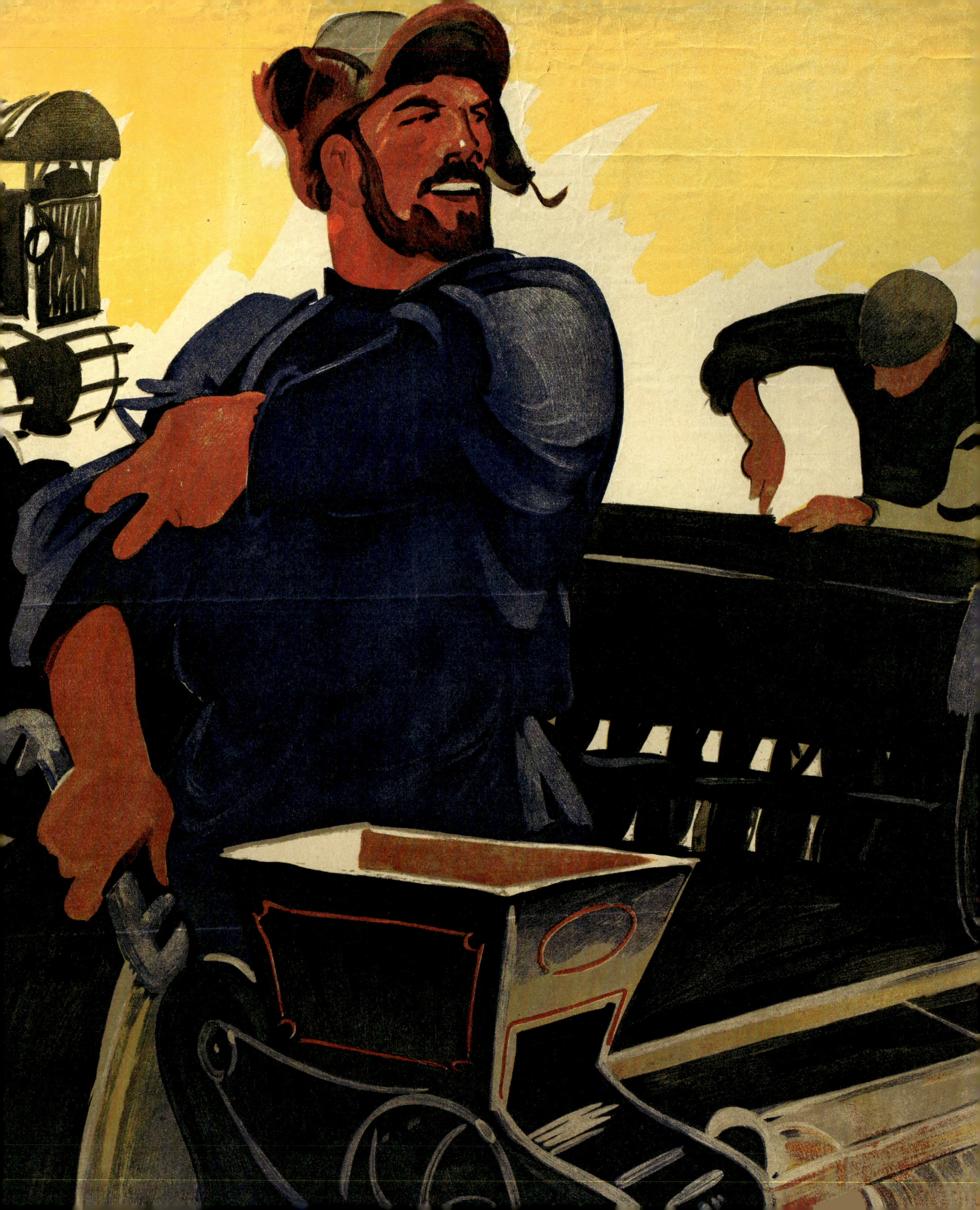

Chapter 2

APOCALYPSE NOW, 1930–1931

The assault was ferocious but helter-skelter with vicious haters of "kulaks" and peasants in general working feverishly to transform the countryside, assert control, and crush resistance. The village did not yield without a fight. Peasants everywhere slaughtered millions of horses and cattle, took up arms, joined rebellions, killed officials, torched government buildings, defended churches, organized protests, distilled grain into moonshine, planted less, refused to cooperate, and curbed their work ethic. The government briefly retreated, then continued the attack.

Throughout December 1929, a Central Committee commission chaired by People's Commissar for Agriculture of the USSR, Iakov Iakovlev (1896–1938), had been busily drafting plans for implementing forced universal collectivization. In actuality, the real decisions were being made by Iagoda, Stalin, and Stalin's closest associates. When Iakovlev submitted his commission's report for review, Stalin sat down with him and transformed a relatively cautious (though still radical) text into a battle plan against the peasantry in general and the "kulaks" in particular. Most important, details of implementation were left out, apparently to unleash the creative violence of local officials.

The finalized decree of January 5, 1930, emphasized economies of scale and dramatic mechanization thanks to the colossal building of farm machinery, but also admitted that such equipment was still in short supply. Therefore, officials should protect draft horses from wanton slaughter. The leadership was right to worry. The OGPU reported on December 22, 1929, and again on January 11, 1930, that the selling off and slaughter of livestock was taking on mass proportions. The later report noted that in one district of Ukraine, 75 percent of livestock had been slaughtered, with 100 percent in some villages.[1] The decree foresaw the completion of collectivization in major grain-producing regions, such as the Lower Volga, the Middle Volga, and the North Caucasus, by fall 1930 or at the latest spring 1931. Since only a tiny proportion of households had been collectivized in these regions, officials would have to work at breakneck speed. Such was presumably the intention. While the decree falsely assumed a powerful and spontaneous grassroots push to collectivize Soviet agriculture, it prudently recommended that party institutions take control and guide it to the end.

The decree's most fundamental goal, because it made possible overall success of the campaign, was the "liquidation of the kulaks as a class." This task had been undertaken with gusto by local officials well before the decree was made public on January 6. In fact, on that very day a central party directive chided party officials for failing to take on "this central task of class warfare in the countryside."[2] This policy was crucial to the success of collectivization for three reasons. First, it removed many of the boldest members of

◀ Detail of "Comrade, prepare in the Bolshevik manner for strenuous sowing!," *see page 53*.

rural communities, the ones most likely to put up a fight. Second, the crushing of the strongest peasants was likely to intimidate those who remained and prevent them from resisting. Third, property confiscated from better-off villagers provided much of the capital upon which to build collective farms.

Local officials got right to work. In the far northern region of Karelia, prosecutors and judicial authorities ordered their subordinates on January 7 to carry out "decisive measures" against "kulaks" and well-to-do elements. Among the characteristics of the targets of attack was ownership of one or two horses or one or two cows—not a very high bar. Central officials were also busy. On January 10, Iagoda ordered regional OGPU bosses in the north, the Urals, Siberia, and the Far East to report on locations for resettlement and the number of "kulaks" who could be settled in their regions. The next day he wrote to his lieutenants that the "kulak class" must be destroyed by March or April.

Iagoda was also scrambling to figure out what to do with the thousands—ultimately millions—of "kulaks" who would soon be flooding into outlying regions. Could they possibly be organized into settlements in the middle of nowhere and put to work without guards? Already back in 1923, Felix Dzerzhinsky (1877–1926), then head of the secret police, had advocated using forced labor for colonizing underdeveloped areas of the country. In 1928, the commissar of justice, Nikolai Janson (1882–1938), proposed using convicts for labor in the northern timber industry. In June 1929, the Politburo approved a resolution, "On the Utilization of the Labor of Criminal Prisoners," to reduce the pressure on overcrowded prisons and economize on their maintenance. Such schemes were a backstory to the unprecedented resettlement of despoiled kulaks starting in winter 1930.

Much of the push for "dekulakization" came from local officials. Party boss Andreev in the North Caucasus requested approval on January 16 for the exile of twenty thousand "kulak" families.[3] With an average of five members per family, that worked out to one hundred thousand people. Similar inquiries came in from other grain-producing regions. It seems, in fact, that many local officials, even down to the village soviet level, had been preparing for such operations for weeks. Pressures from above and below clearly reinforced each other. On January 25, Moscow ordered local soviets "to mobilize broad landless, poor, and middle peasant masses for the reorganization of rural life on socialist foundations."[4] For such operations, secret police troops and officers, militia, and squads of factory workers were armed and sent into the countryside. The London *Times* reported that officials were aiming to "stamp out kulaks entirely" and had sentenced to death in one locality twelve peasants charged with "agitating against the rural policy and attempting to discredit Communists in the eyes of the masses."[5]

The Politburo finally got around to spelling out the rules for "dekulakization" in areas of universal collectivization on January 30. The plan had been worked out by a committee headed by Molotov. The plan was to apply the policy to 3–5 percent of all peasant households. In reality, according to official figures, by the end of 1929 only 2.2 percent of households could be so categorized.[6] Surely the higher number was put out to give officials greater leeway in punishing those who resisted.

Those defined at the local level as "kulaks" were to be divided into three categories, in declining order of dangerousness. Those in category one were to be sent to concentration camps or executed; in category two, exiled to distant regions; and in category three, resettled regionally. None was to be admitted into collective farms. All of their property,

Fig. 19 N. B. Terpsikhorov. "Join the collective farm." 1930. "Let's give the kulak a united rebuff. Collectivize your livestock. With all your inventory join the collective farm, don't slaughter or sell your animals."

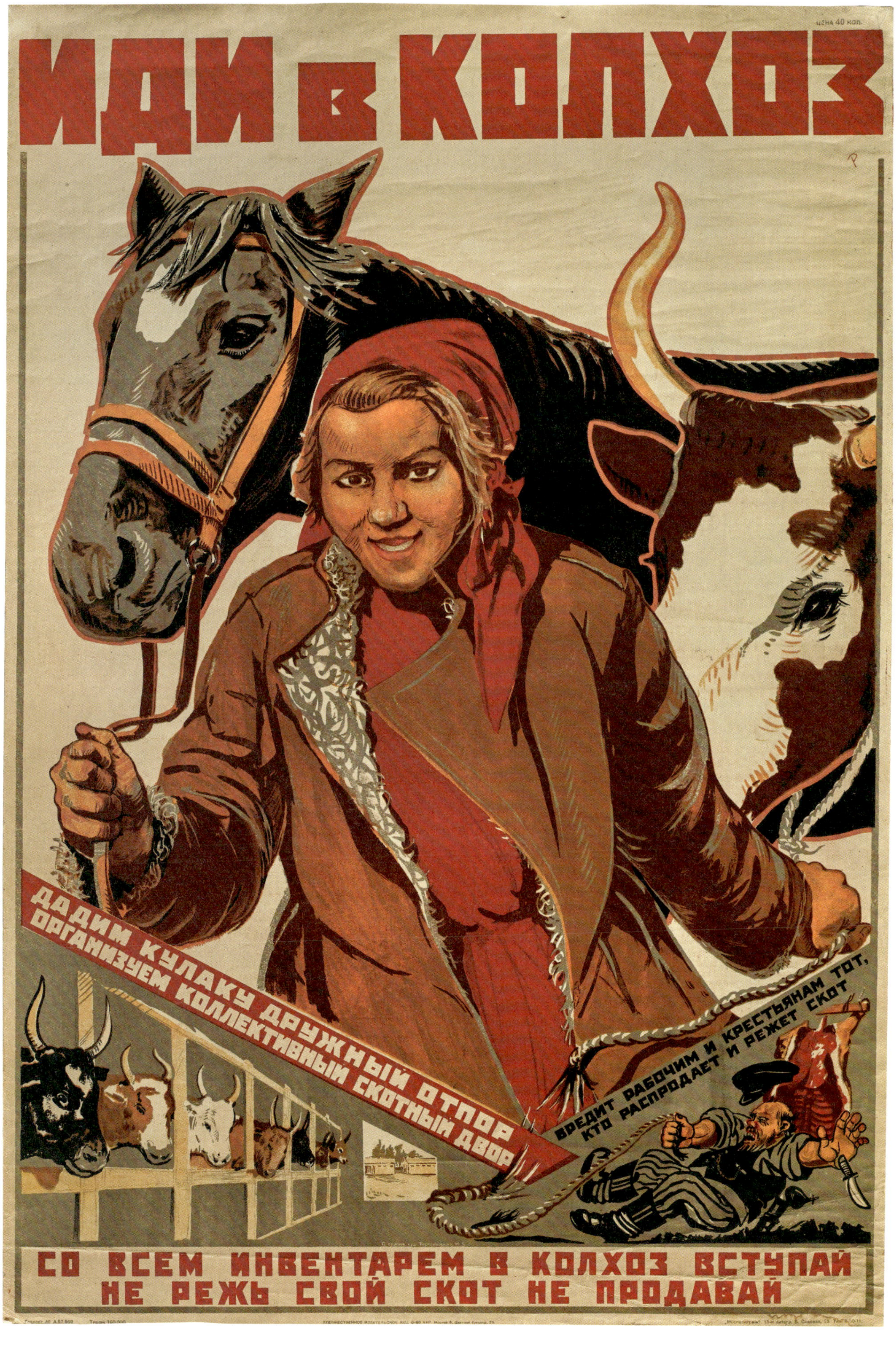
цена 40 коп.
ИДИ в КОЛХОЗ
ДАДИМ КУЛАКУ ДРУЖНЫЙ ОТПОР
ОРГАНИЗУЕМ КОЛЛЕКТИВНЫЙ СКОТНЫЙ ДВОР
ВРЕДИТ РАБОЧИМ И КРЕСТЬЯНАМ ТОТ,
КТО РАСПРОДАЕТ И РЕЖЕТ СКОТ
СО ВСЕМ ИНВЕНТАРЕМ В КОЛХОЗ ВСТУПАЙ
НЕ РЕЖЬ СВОЙ СКОТ НЕ ПРОДАВАЙ

Цена 40 коп.

СОВЕТЫ ЛИЦОМ К КОЛХОЗНОМУ ДВИЖЕНИЮ

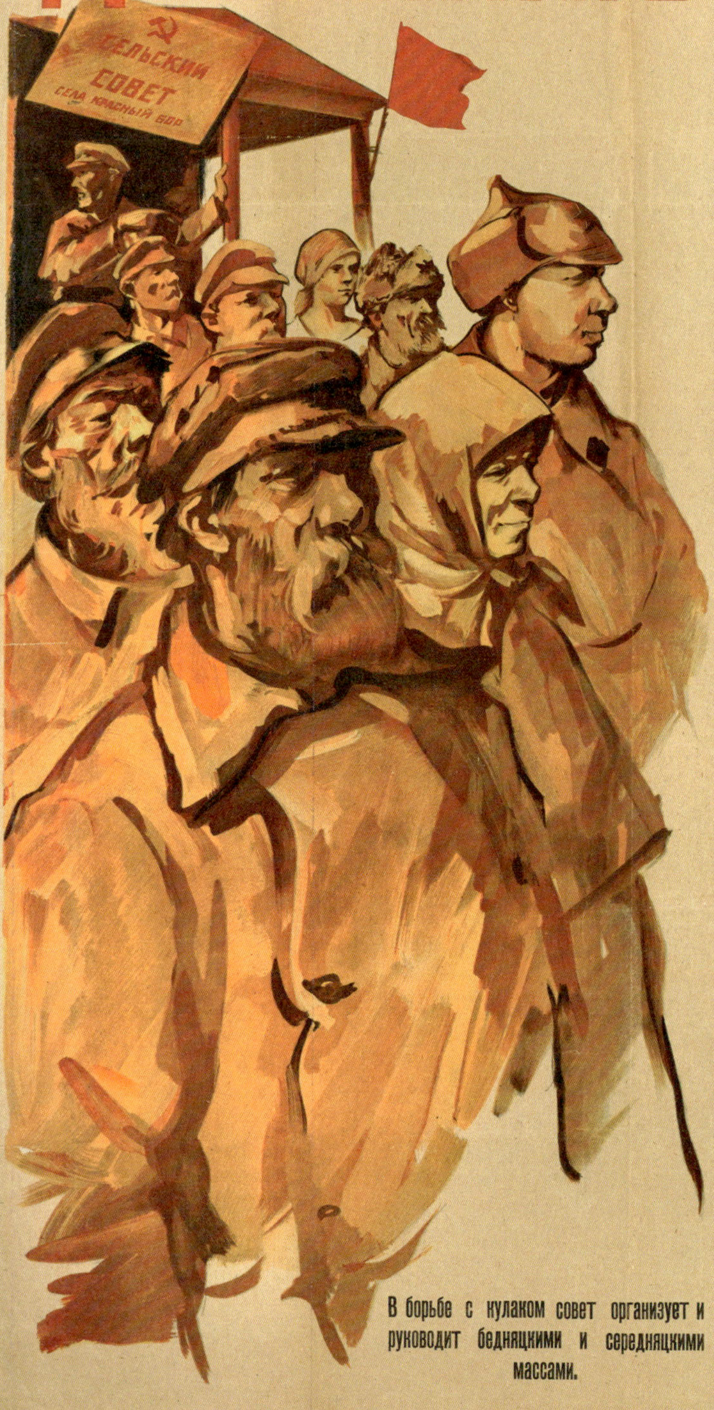

В борьбе с кулаком совет организует и руководит бедняцкими и середняцкими массами.

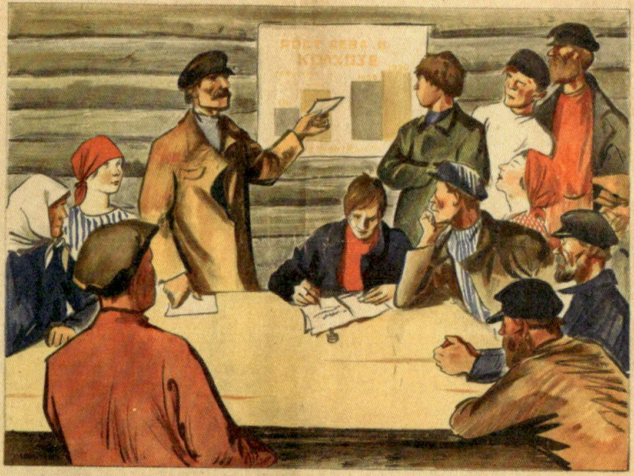

Совет помогает колхозу проводить производственный план.

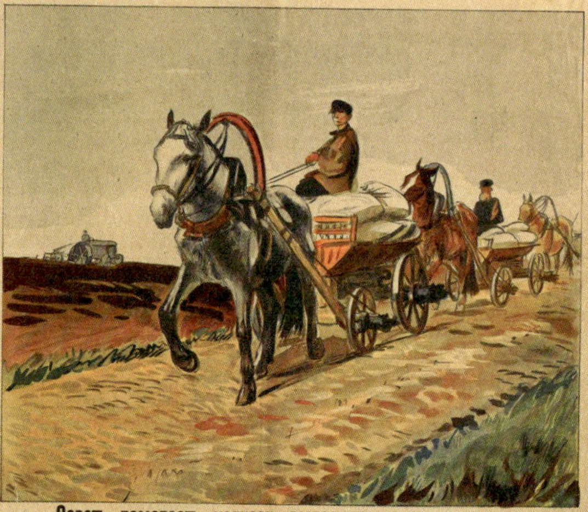

Совет помогает колхозу выполнять государственные обязательства, использовать земельные фонды, своевременно получать кредит.

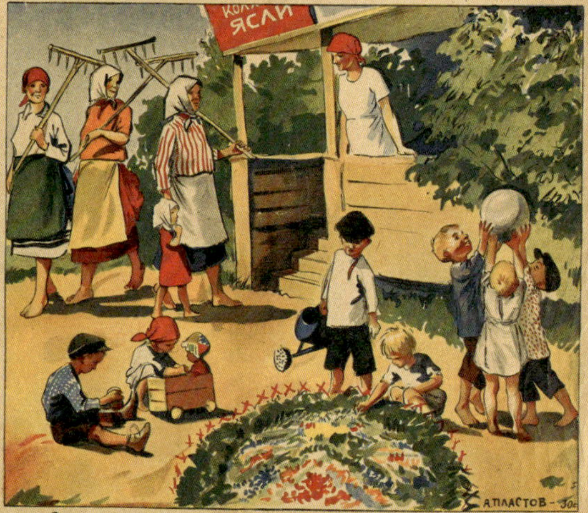

Совет помогает культурному строительству в колхозе.

ГЛАВЛИТ А—49719. Р. 30. № 93. ГИЗ 40434. ЗАКАЗ 1210. ТИРАЖ 35 000.

ГОСУДАРСТВЕННОЕ ИЗДАТЕЛЬСТВО
МОСКВА 1930 ЛЕНИНГРАД

Типо-литография Госиздата «Красный пролетарий», Москва.

Fig. 20 A.A. Plastov. "Soviets, support the collective farm movement." 1930. "The Soviet helps the collective farm to carry out the production plan. The Soviet helps the collective farm to fulfill the state obligations, to use the land funds, and to receive a credit in a timely manner. The Soviet helps the collective farm in its cultural development. In the fight with the kulak, the Soviet organizes and manages the poor and middle peasant masses."

aside from restricted amounts of tools and personal effects, was to be confiscated. Astonishingly, at least half of the exiles were expected to be resettled by April 15, even though, technically speaking, the operation had not even begun. Nor had it been decided where they would be sent or how. Everything was worked out on the fly without attention to legal niceties. The Bolshevik Decree on Land, one of the first statutes adopted after the seizure of power in 1917, which guaranteed the right to work the land with one's own labor, was neither mentioned nor rescinded. De facto, it simply lost force. Nor did the policy remain confined to regions of "universal collectivization": it was implemented everywhere.

Over the next few days, a flurry of orders went out. One charged various officials with drafting timetables and selecting destinations. Another directed officials to prevent the sale or slaughter of "kulaks'" property and to transfer the plunder first to collective and state farms and second to the poor. Yet another specified which family members had to resettle with "dekulakized" heads of household (all of them—even children and old people). Another still required deploying tens of thousands of auxiliary forces to help implement the policy (secret police troops, party members, Komsomol, and factory workers)—but absolutely no regular military personnel (even their families should not be touched, whatever their socioeconomic status). Finally, a propaganda campaign should saturate the news media, and an antireligious campaign should accompany the operation.

It might seem surprising that the party elites would take on not only the entire peasantry and its leaders but also religious believers at the same time. But in the Bolshevik outlook, they were mutually reinforcing or even one and the same. In fact, many reports listed rural church authorities among the "dekulakized." The faithful in general were more likely to oppose socialization than were the secular. Transforming the world in the image of Marx's ideals presupposed absolute dedication and an unwillingness to divide loyalties. Religious faith got in the way. Thus, an antireligious campaign was waged across the country in tandem with collectivization. It involved banning religious processions, mass burnings of icons, staging mock executions of religious symbols, taking down church bells, and closing churches. Such actions were part of what scholars have called a "cultural revolution" that broke out during the years of the first Five-Year Plan. As of March 1, officials had counted 6,715 shuttered churches.[7] Among the most potent rural symbols of religious belief and practice was the church bell, which not only called the people to worship but could alert them to danger. A confiscation campaign, often instigated by the League of the Militant Godless, commenced in October 1929 and by the end of January had yielded nearly 700 tons of metal in the Central Black Earth Region alone.[8] Such actions provoked hundreds of mass disturbances in rural areas, especially among women.

Yet closing churches, melting down bells, and exiling and killing priests did not immediately cause religious faith and practice to recede in the Russian countryside. A Bolshevik official deeply involved with antireligious activism, Emelian Yaroslavsky (1878–1943), reckoned that fewer than 10 million out of 130 million people in the countryside had broken with religion as of 1930.[9]

The deportations of "kulaks" began in February in the dead of winter. In the far north, five trainloads arrived daily from mid-February, each jammed with five thousand to seven thousand human beings. Families had the right to take two

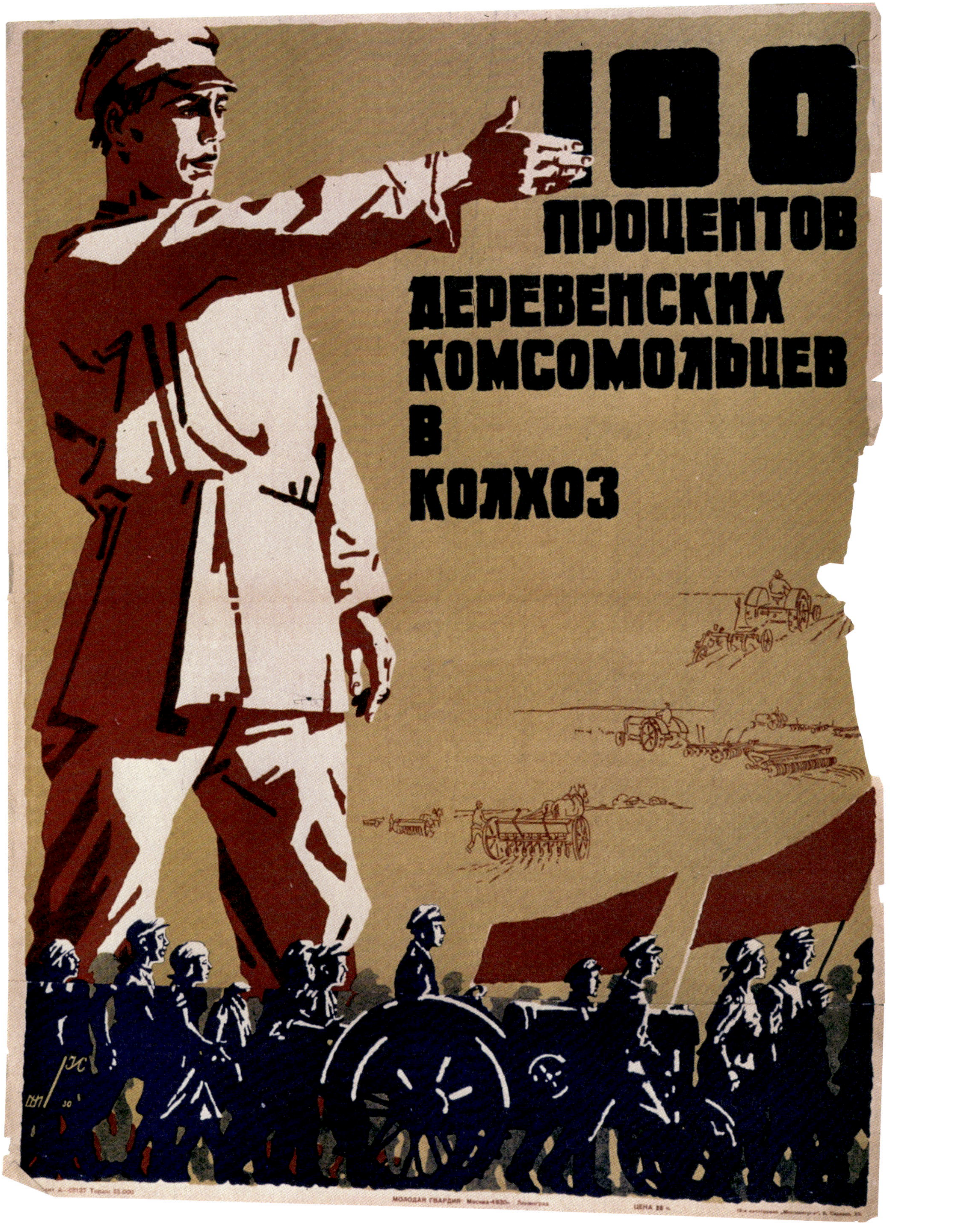
100
ПРОЦЕНТОВ
ДЕРЕВЕНСКИХ
КОМСОМОЛЬЦЕВ
В
КОЛХОЗ
МОЛОДАЯ ГВАРДИЯ Москва-1930- Ленинград
ЦЕНА 25 к.

◀ **Fig. 21** P. Gorily (Gorelyi). "100% of rural Komsomol members into the collective farm." 1932.

▶ **Fig. 22** M.M. Cheremnykh. "Peasant women, join the collective farm!" 1930.

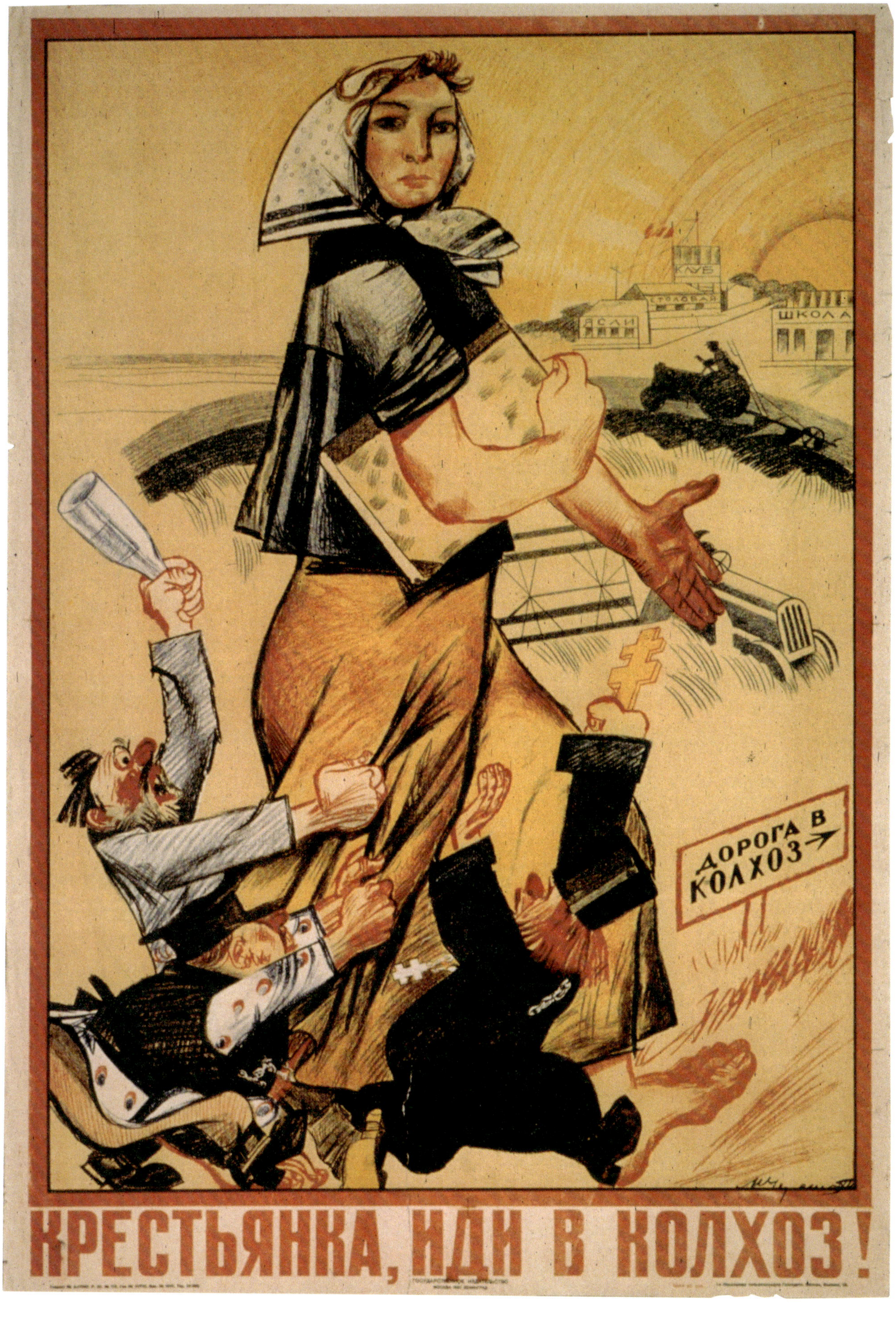

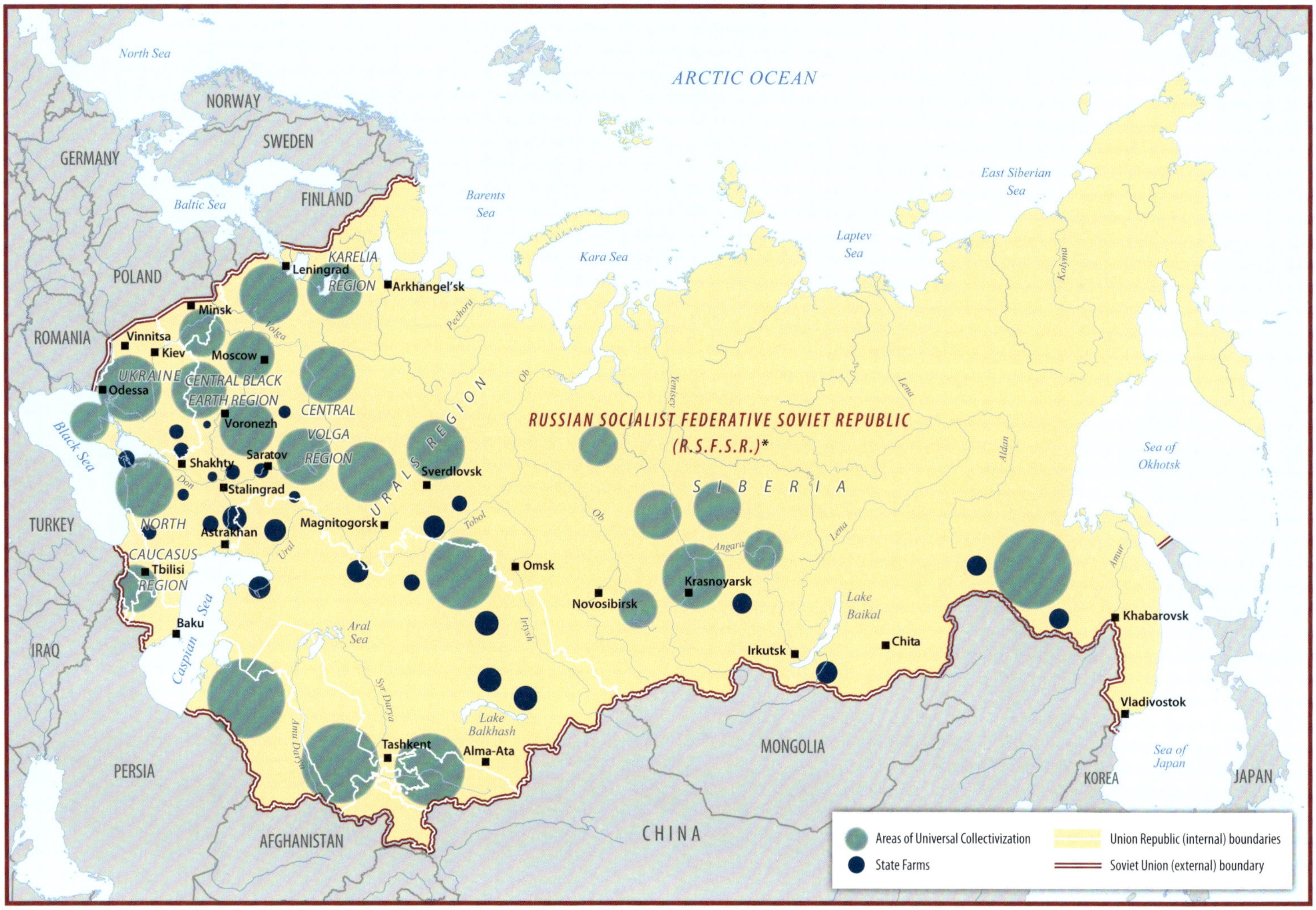

** Country name changed to Russian Soviet Federative Socialist Republic in 1937.*

Map 2 Areas of universal collectivization and state farms months' worth of food, as well as 1,000 pounds (30 poods) of equipment and personal belongings, but few did. Tens of thousands were dumped in the middle of nowhere. Many were unfit for work. Those who could work often had to walk dozens or even hundreds of miles to their work settlements, where, in addition to fulfilling unconscionably high lumber quotas, they also had to build their own housing—usually without adequate tools or skills—so their families could join them in the middle of the forest. And the convoys kept arriving. Service personnel were promised—including doctors, veterinarians, teachers, and agronomists—but often were slow in coming.

The reasons for being "dekulakized" varied. Living in a brick house was sufficient. Displaying religious icons in one's home also. Many worker and party activists were fanatics who terrorized villagers. Secret police officials with quotas to fill often barged into homes in the dead of night, competing with one another. The victims were not seldom shamed and humiliated, in addition to being despoiled and exiled. The police did not always enjoy the support of rural officials, who often preferred to protect their own. Even so, as of February 15, according to official statistics, 64,589 people were in the process of being "dekulakized."[10]

Because the operation was conducted in such a haphazard manner, many of the intended procedures were violated. Peasants continued to sell or slaughter their livestock, causing huge problems in agricultural production for years to come. Since local officials often carried out the policy without secret police supervision, many "kulaks" managed to flee to urban areas, where they "wormed their way" into construction sites, factories, and other places of employment, much to the consternation of senior officials, who typically viewed them as irredeemably "dangerous social elements." The families of Red Army men not seldom suffered dispossession and exile.

Given the systematic outrages committed against millions of peasants, the regime not surprisingly had to contend with mass unrest and resistance, both violent and passive. According to official statistics, throughout 1930 such events in the countryside totaled 13,754, with over 3.3 million participants, the largest share concentrated in February through April. Such numbers dwarfed those experienced during the revolution of 1905–6 and involved far more violent attacks against people (representatives of the regime) than against property (landed estates), though widespread arson also occurred.[11] Moreover, anti-Soviet and anti-Communist rhetoric pervaded the countryside, along with images of impending war, famine, and a return to serfdom. Many of the participants in rural unrest were women. It was typical for a crowd of several hundred to gather in a village to protest collectivization, the closing of churches, the despoiling and exile of fellow villagers, and other deeply unpopular acts. Women often took prominence in the protest movements, especially in March 1930, because it was believed that the authorities would deal less harshly with them. (It is less clear why their prominence in such actions tapered off after March.)

The widespread protests, mass slaughter of livestock, and reports of looming famine caused the leadership to back off. As a secret Central Committee report of April 2 admitted, "If measures had not immediately been taken against the distortion of the party line, we would have had a broad wave of insurrectionary peasant outbursts, a good half of our 'lower-level' officials would have been slaughtered by peasants, sowing would have been disrupted, collective farm construction would have been destabilized, and our internal and

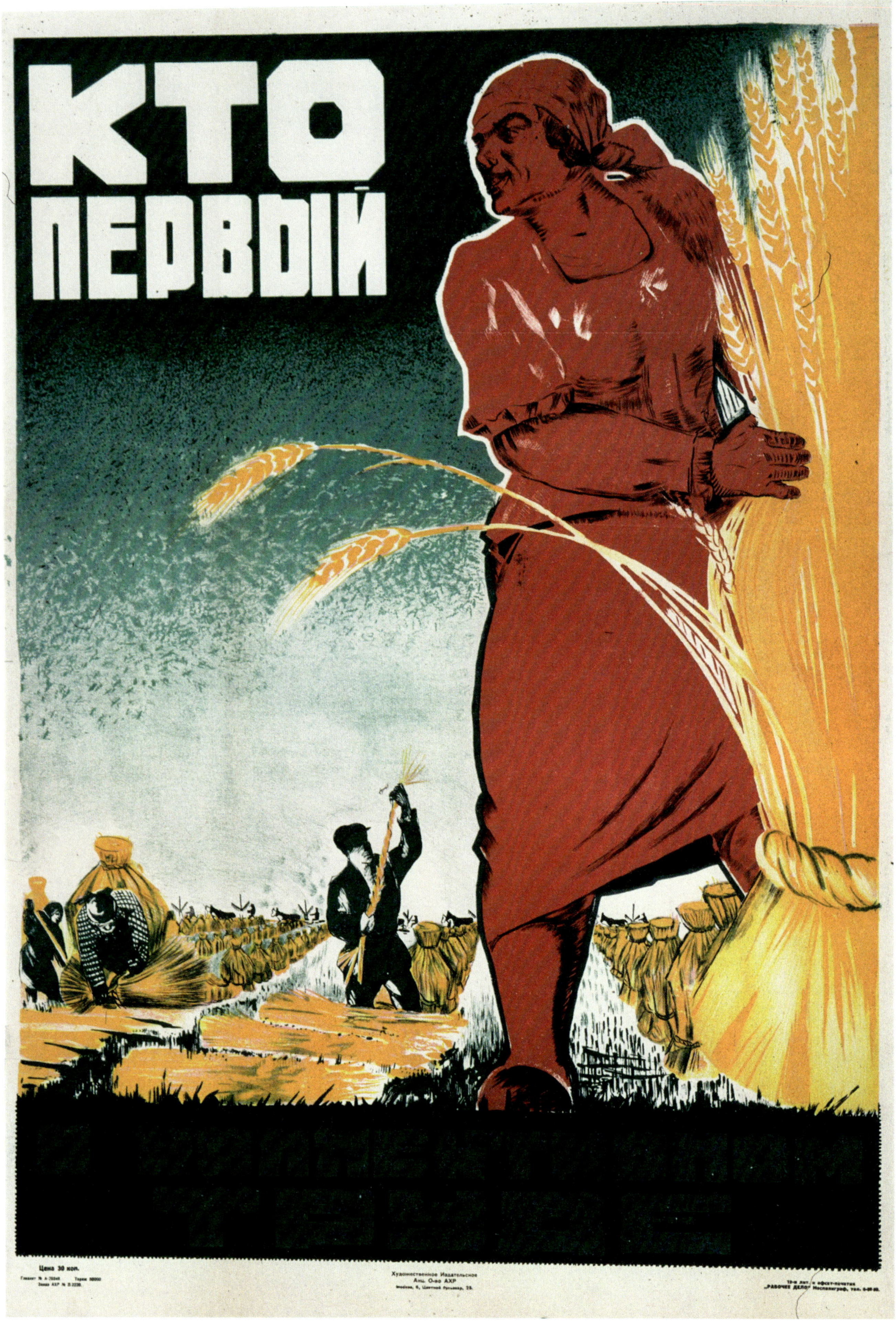
КТО
ПЕРВЫЙ
Цена 30 коп.
Художественное Издательское
Акц. О-во АХР

Fig. 23 Artist not identified. "Who is the first in the collective labor?" 1928–32?

external situation would have been placed in jeopardy."[12] Stalin's famous "Dizzy with Success" article in the March 2 issue of *Pravda* was the official response to these threats. It was a tour de force of hypocrisy. He claimed that collectivization was meant to be entirely voluntary and that any deviations from this approach were to be blamed entirely on local officials. Naturally, they felt profoundly slighted. To read the article, one would think Stalin a moderate on cultural revolution, mocking the "revolutionaries" who imagined that the appropriate way to organize collective farms was first and foremost "by removing the bells from the churches. Just imagine, removing the church bells—how r-r-revolutionary!"[13]

Most peasants breathed a sigh of relief, though many felt confused, betrayed, or suspicious. Millions immediately tore up their contracts to enter collective farms. Not surprisingly, squabbles over retrieving socialized property went on for months. Stalin's speech seems to have aroused more than quieted the peasantry: 6,528 of the year's 13,754 rural disturbances took place in March, almost as if Stalin's apparent leniency suggested reprisals would be mild. But of course Stalin had not meant it. The disturbances were crushed fiercely. True, a Central Committee directive of March 14 asserted that joining a collective farm was a strictly voluntary act, churches should be closed only when the majority of villagers clamored for it, and officials who abused the religious feelings of peasants had to be prosecuted. But few officials actually suffered punishment. Rather, higher-ups simply encouraged them to conduct better preparatory work going forward. In other words, collectivization and "dekulakization" would continue. Still, there was a lull through the summer, and sowing took place successfully across the country. Rising hunger probably pushed many to sow.

Backing off from collectivization did not mean sending the 46,261 exiled "kulak" families (or roughly 230,000 individuals) back to their homes. They were either on their way to distant and sparsely inhabited territories in the far north (convoys kept arriving well into the summer) or had already been dumped there. On April 5, 1930, the Politburo created a commission to look into the 35,000 affidavits of false accusation submitted by the families. It reviewed 23,360 cases, concluded that 10 percent should not have been exiled, and proposed returning 2–3 percent.[14] The Politburo allowed only army veterans and their families to return home. In other words, almost none.

An order of the agriculture ministry of the Russian Republic of April 1, 1930, went into detail about what the settlements should look like. They should contain at least twenty but not more than one hundred households. A wide variety of services and resources—medical, educational, retail, agricultural supply—should be provided, though in practice many such settlements scraped by for months and even years with few such amenities. Agricultural land available to them should be of poor quality. No individual farms were allowed. The next day the committee on the liquidation of "kulaks" of the Republic of Karelia spelled out details of their campaign. Lists of peasants to be "liquidated" should be determined by a troika on the basis of decisions by the local population. It was essential, according to the committee, that genuine "kulak" households be affected, not the merely well-to-do or middle peasants or those stripped of voting rights, like former policemen and priests. Of course, since the definition of *kulak* was both vague and extremely capacious, it is hard to believe that simple villagers could manage such fine distinctions. Often even local secret police officials were not much better

Притча о Справедливости.

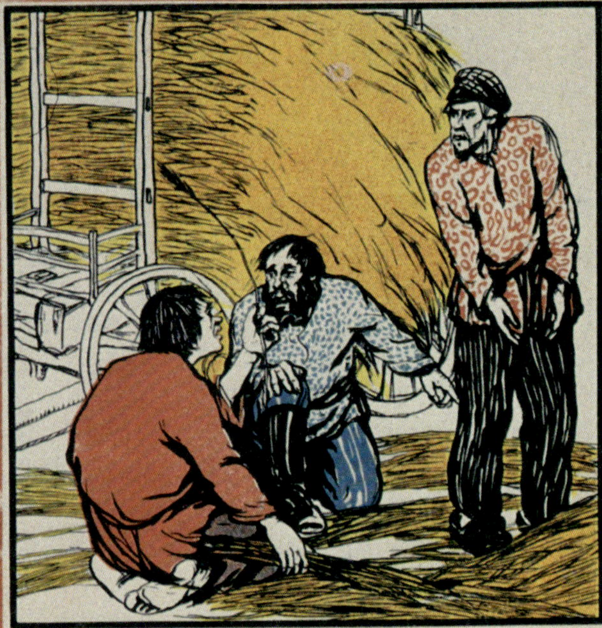

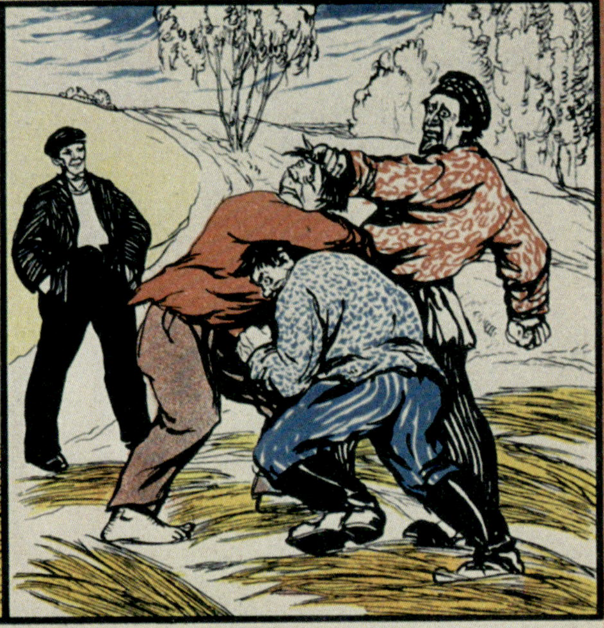

У отца было три сына, три круглых дурака.

Умер отец и оставил на дурацкое горе наследство: воз соломы.

Стали дураки судить-рядить, как им наследство поделить.

Дураки эти были справедливые. Решили поделить так, чтобы ни один не получил больше другого ни на былиночку...

Сначала делили в раскладку: одну соло- [illegible]ому, другую другому. [illegible] несправедливость. Увидели, что тут большая [illegible] соломинка длиннее, другая короче...

Тогда решили [illegible] разделить на три сорта: на длинную, среднюю и [illegible], а потом и произвести раскладку по дурацким душам.

Но и тут справедливости не вышло. Когда распределили солому по длине, то в глаза бросилась другая разница: которая соломинка оказалась толще, которая тоньше...

Пришлось каждый сорт делить еще на три сорта: на толстую, среднюю и тонкую солому. Получилось 9 сортов.

Приступили к раскладке. А солома-то оказалась все-таки разная: которая с комлем и с вершинкой, другая с комлем, но без вершинки, третья наоборот, а четвертая без того и без другого...

Для справедливости пришлось каждый из 9 сортов делить еще на 4. Уже оказалось 36 сортов. Каждый сорт надо разнести на [illegible]

А тут [illegible] получалось 108 долей.

[illegible] горе: среди соломы дураки вдруг нашли 2 необмолоченных колоса.. Задача: как их разделить на 3 части?

Думали-думали, даже вспотели... Ничего не придумали, и подрались. Уставши от драки, решили колосья разделить не по справедливости, а по жеребью. И опять принялись за раскладку.

И вот ходят от кучки до кучки и все раскладывают,—устанавливают справедливость...

День сменился вечером, прошла ночь, наступило утро, а они все раскладывают... Но до справедливости еще далеко...

Подошел тут прохожий и спрашивает: в чем дело?

Рассказали ему дураки про свое горе: сколько они трудов положили на эту самую справедливость, которой конца-краю не видно.

— А что вы с соломой потом будете делать? — спросил прохожий.

— В город, на базар, продавать повезем,—отвечали дураки.

— Эх вы, дураки, дураки! Охота вам такую канитель разводить. Да вы бы солому продали целиком, а делить бы стали деньги!..

Но дураки были в таком азарте от дележки, что и слушать ничего не хотели.

Так они до сей поры и делятся. Топчутся вокруг своих соломенных долей, строго наблюдая, чтоб кому не перепала лишняя былиночка.

Точно на роду дуракам написано: весь век делиться по справедливости, и никак не разделиться...

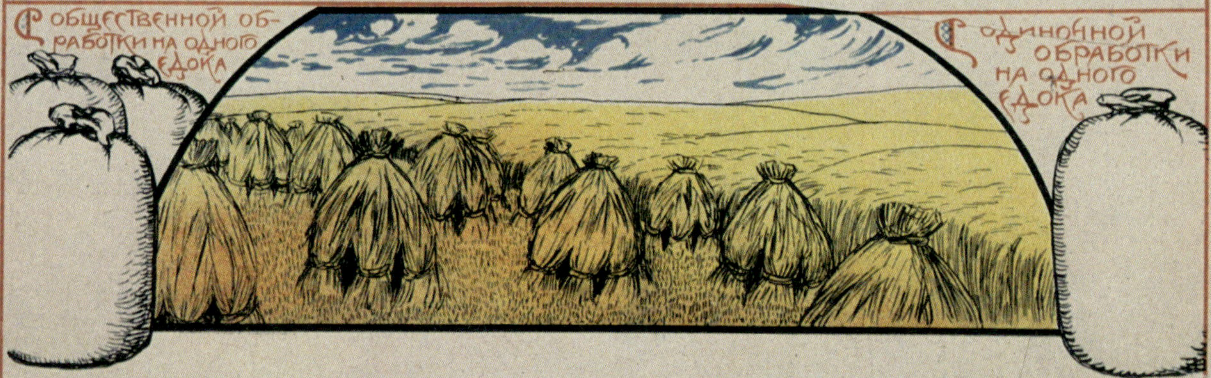

Многие спросят: к чему такая притча? Неужели есть на свете подобные дураки?..

Есть. И очень много.

Те из крестьян, которые до сей поры цепляются за старый способ распределения земли, поступают точь-в-точь, как эти дураки в притче.

Если нельзя разделить солому поровну, то нельзя разделить землю, которая никогда не бывает одинаковой по качеству. И все-таки во многих местах России земля делится даже по более капительному способу, чем в этой притче.

И вот получаются бесчисленные полоски на пространстве многих верст. Переезд с места на место занимает больше времени, чем обработка земли. Полоски «справедливые», но настолько узкие, что не только лошадь с плугом, а и человек не повернется на ней. А сколько пропадает земли под межами! А сколько греха, ссоры и драк при самом разделе!..

Тут справедливости мало. Жадность, зависть, боязнь, чтобы сосед не получил больше—вот что тут, а совсем не справедливость.

Но как же быть? Ведь нужно же так или иначе распределить землю между работниками?

Ответ на это, как и в притче:

— Зачем делить землю, когда гораздо легче разделить урожай!..

— А сделать это можно с помощью общественной обработки земли.

Государственное Издательство

2-я Государствен. типография.

◀ **Fig. 24** Artist not identified. "A parable of fairness." 1920. Sons fighting over inherited hay versus collective ownership.

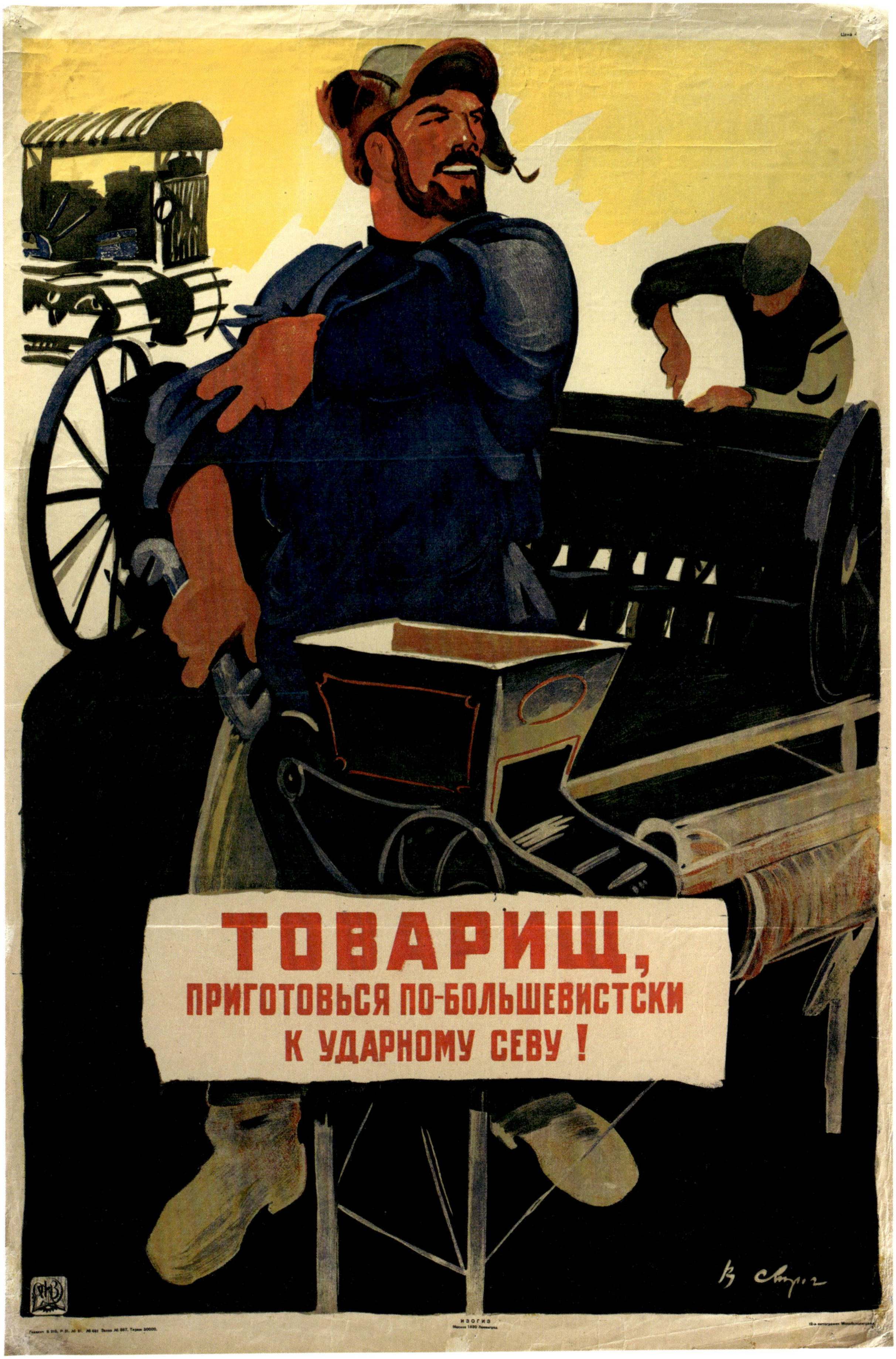

▶ **Fig. 25** V.S. Svarog (Korochkin). "Comrade, prepare in the Bolshevik manner for strenuous sowing!" 1930.

prepared; some could not sign their names. So, to prevent "excesses" (*peregiby*, meaning violence and abuse), on April 2 the Central Committee forbade sending them into areas of peasant unrest without the approval of secret police authorities in Moscow. In fact, despite—or in part because of—"madly overdetailed plans, blueprints, and building specifications," according to Lynne Viola, the most knowledgeable Western expert on Soviet collectivization, the entire grandiose project was a "bureaucratic nightmare."[15]

Iagoda was starting to figure out how to wring a profit from the chaos. In a memorandum of April 12, 1930, he advocated transforming the settlements into "*colonization villages* [*kolonizatsionnye poselki*] without any expectation of a set period of imprisonment."[16] His idea was to settle the far north with convicts and set them to work in the logging industry. But crash industrialization was under way, and other authorities wanted slave labor, too, if it was available. At the end of the month, for example, the labor department in the Ural region reported that local industry was looking to increase its labor force by nearly 450,000. Later in the summer (August 18), the leadership decreed the creation of special colonies in the Ural region and in Siberia. By the end of 1930, there were 250 such camps, 35 of them in the Ural region, where up to 25 percent of all industrial workers had been repressed, mostly as "kulaks."[17] Logging, mining, and construction thus became the main occupations of what formerly had been the country's most skilled farmers.

Here one sees the big difference between 1930–31 and 1937–38, between collectivization and "dekulakization," on the one hand, and the Great Terror, on the other. In the first nine months of 1930, the secret police summarily executed 18,966 "kulaks"—whereas in the course of 1937 and 1938, they slaughtered a minimum of 681,692 people. During Stalin's first mass operation, removing "alien elements" mattered less and generating slave labor more, compared to his second one.

Industrial slave labor was undoubtedly especially welcome at a time when the deepening Great Depression was driving down commodity prices worldwide. A bushel of wheat in Liverpool, for example, had fetched between $1.30 and $1.50 in 1927 and 1928, but only 58⅝ cents by the end of 1930, the lowest price since 1894. Since the cost of machinery did not fall, the Soviets now needed to export two or even three times more grain to acquire the industrial equipment anticipated by the Five-Year Plan. Throughout the summer, activists fanned out into the countryside "to seize the very last grain so that the kulaks could not use the bony hand of famine to choke the Soviet government," as one plenipotentiary recalled from the official preparatory rhetoric.[18] Sadly, official pressure to collect the grain quickly and dispatch it immediately from the countryside resulted in significant wastage, with as much as 22 percent of the grain collected that summer falling to the ground or rotting.[19]

The squeeze on resources needed for industrialization meant that the "special settlements" were not prioritized with the huge investments needed to make them livable. The first makeshift dwellings, which in many cases took a year or two to complete (in one northern district, only 3 percent of housing had been built as of mid-September), often had dirt floors, no chimneys, open-dormitory space division, bunk beds, and no window screens to protect against the swarms of voracious northern biting insects. Medical care was abysmal. As of April 11, 1930, there was roughly one doctor for every ten thousand special settlers in the far north. The following year, the only pediatrician for the

Fig. 26 A. M. Niurenberg. Okno IZOGIZa. Plakat-gazeta № 12. 1931. Conscientious poor and middle peasants working hard on the collective farm contrasted with a conniving and lazy kulak who joins the farm and then disappears.

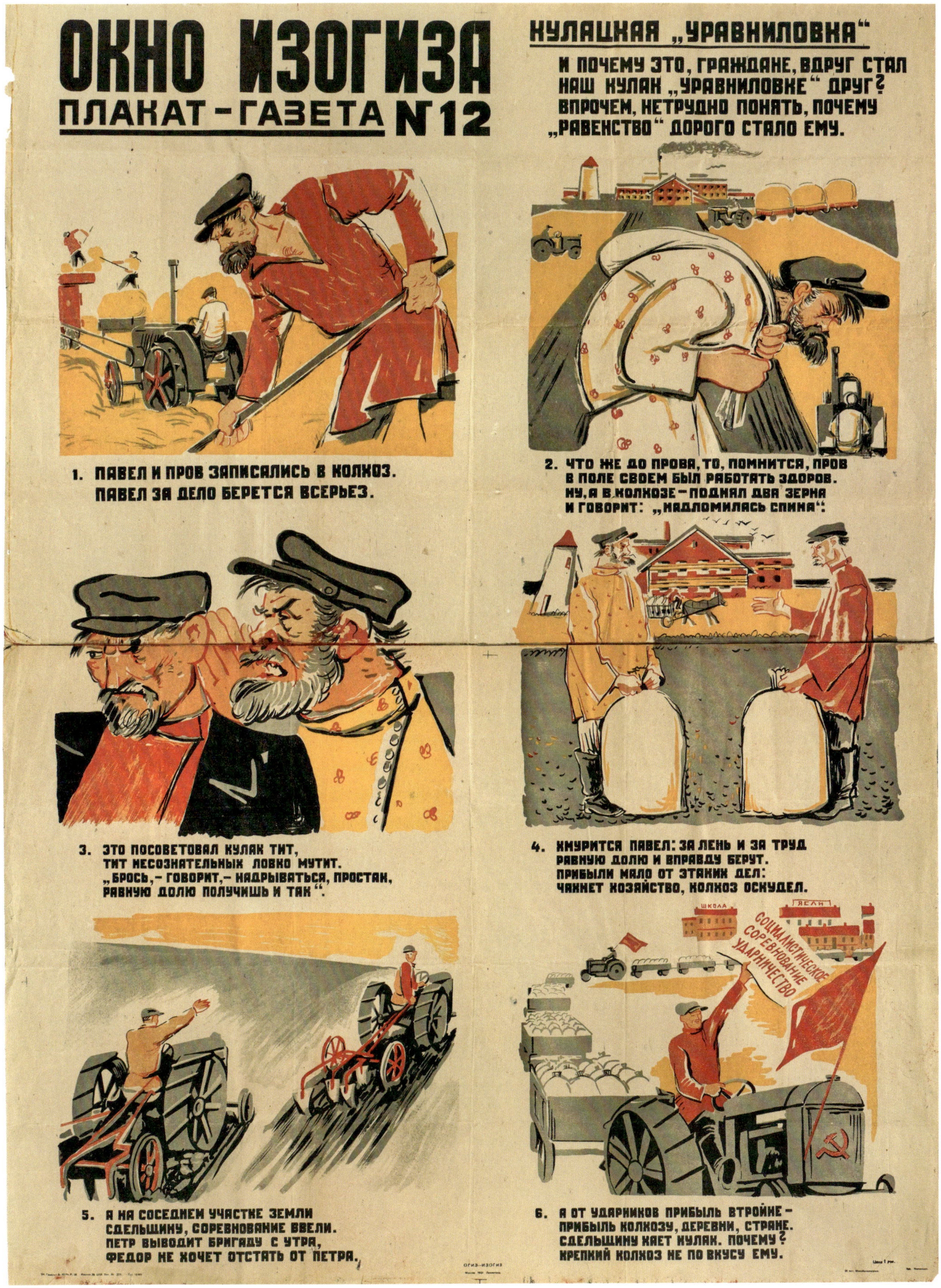
ОКНО ИЗОГИЗА
ПЛАКАТ – ГАЗЕТА №12
КУЛАЦКАЯ „УРАВНИЛОВКА“
И ПОЧЕМУ ЭТО, ГРАЖДАНЕ, ВДРУГ СТАЛ НАШ КУЛАК „УРАВНИЛОВКЕ“ ДРУГ? ВПРОЧЕМ, НЕТРУДНО ПОНЯТЬ, ПОЧЕМУ „РАВЕНСТВО“ ДОРОГО СТАЛО ЕМУ.
1. ПАВЕЛ И ПРОВ ЗАПИСАЛИСЬ В КОЛХОЗ. ПАВЕЛ ЗА ДЕЛО БЕРЕТСЯ ВСЕРЬЕЗ.
2. ЧТО ЖЕ ДО ПРОВА, ТО, ПОМНИТСЯ, ПРОВ В ПОЛЕ СВОЕМ БЫЛ РАБОТАТЬ ЗДОРОВ. НУ, А В КОЛХОЗЕ – ПОДНЯЛ ДВА ЗЕРНА И ГОВОРИТ: „НАДЛОМИЛАСЬ СПИНА“.
3. ЭТО ПОСОВЕТОВАЛ КУЛАК ТИТ, ТИТ НЕСОЗНАТЕЛЬНЫХ ЛОВКО МУТИТ. „БРОСЬ, – ГОВОРИТ, – НАДРЫВАТЬСЯ, ПРОСТАК, РАВНУЮ ДОЛЮ ПОЛУЧИШЬ И ТАК“.
4. ХМУРИТСЯ ПАВЕЛ: ЗА ЛЕНЬ И ЗА ТРУД РАВНУЮ ДОЛЮ И ВПРАВДУ БЕРУТ. ПРИБЫЛИ МАЛО ОТ ЭТАКИХ ДЕЛ: ЧАХНЕТ ХОЗЯЙСТВО, КОЛХОЗ ОСКУДЕЛ.
5. А НА СОСЕДНЕМ УЧАСТКЕ ЗЕМЛИ СДЕЛЬЩИНУ, СОРЕВНОВАНИЕ ВВЕЛИ. ПЕТР ВЫВОДИТ БРИГАДУ С УТРА, ФЕДОР НЕ ХОЧЕТ ОТСТАТЬ ОТ ПЕТРА.
ШКОЛА
ЯСЛИ
СОЦИАЛИСТИЧЕСКОЕ СОРЕВНОВАНИЕ УДАРНИЧЕСТВО
6. А ОТ УДАРНИКОВ ПРИБЫЛЬ ВТРОЙНЕ – ПРИБЫЛЬ КОЛХОЗУ, ДЕРЕВНИ, СТРАНЕ. СДЕЛЬЩИНУ ХАЕТ КУЛАК. ПОЧЕМУ? КРЕПКИЙ КОЛХОЗ НЕ ПО ВКУСУ ЕМУ.
ОГИЗ–ИЗОГИЗ
Цена 1 руб.

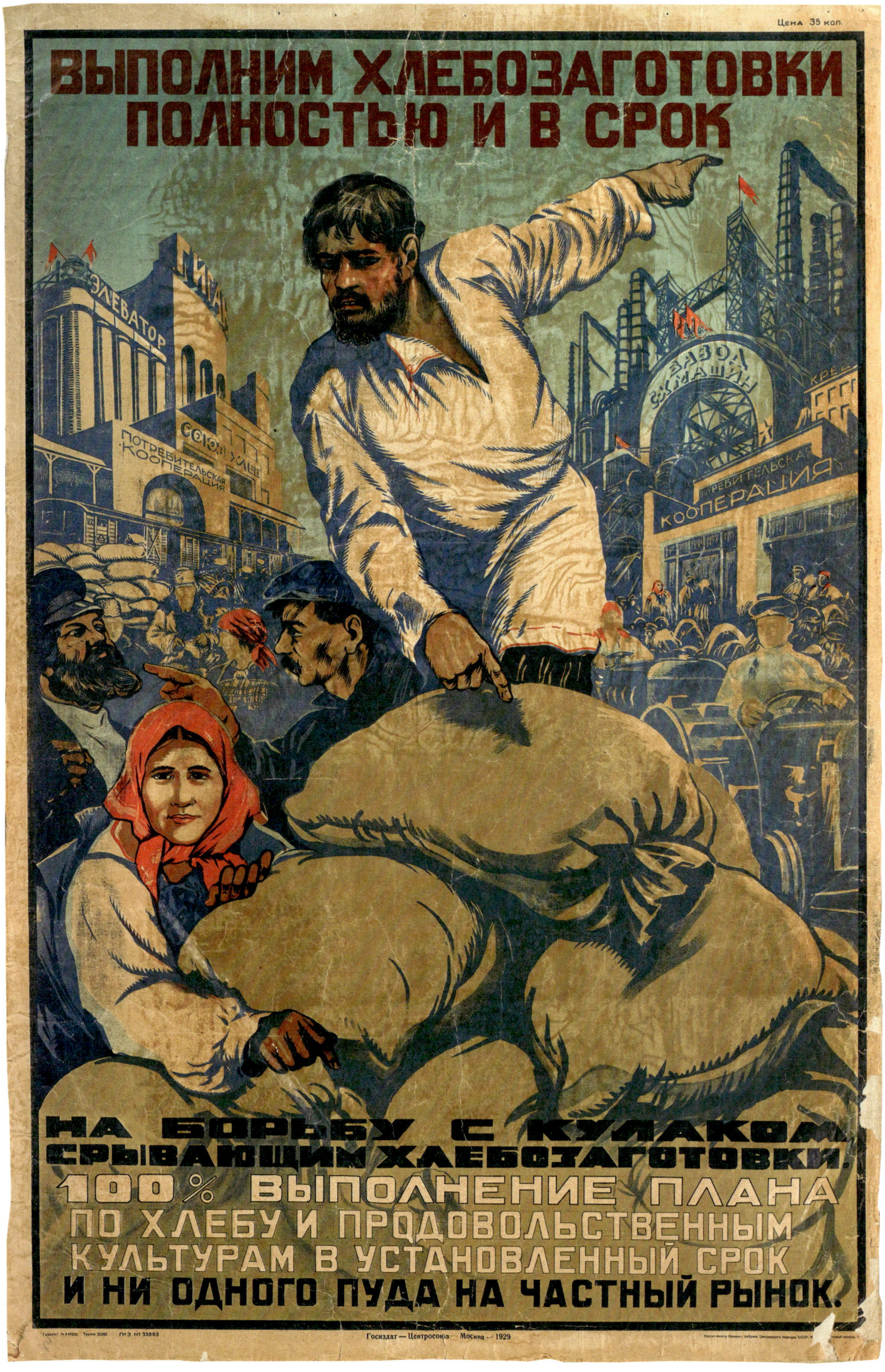
Цена 35 коп.
ВЫПОЛНИМ ХЛЕБОЗАГОТОВКИ
ПОЛНОСТЬЮ И В СРОК
ЭЛЕВАТОР
ПОТРЕБИТЕЛЬСКАЯ КООПЕРАЦИЯ
ЗАВОД С/Х МАШИН
КООПЕРАЦИЯ
НА БОРЬБУ С КУЛАКОМ
СРЫВАЮЩИМ ХЛЕБОЗАГОТОВКИ.
100% ВЫПОЛНЕНИЕ ПЛАНА
ПО ХЛЕБУ И ПРОДОВОЛЬСТВЕННЫМ
КУЛЬТУРАМ В УСТАНОВЛЕННЫЙ СРОК
И НИ ОДНОГО ПУДА НА ЧАСТНЫЙ РЫНОК.
ГИЗ № 33963
Госиздат — Центросоюз Москва — 1929

vast Magnitogorsk metallurgical complex, where huge numbers of exiled "kulaks" were employed, wrote in an official report:

> When you walk through the barracks, everywhere, at every step you bump up against children with diarrhea, uncontrolled bleeding, inflammation of the lungs. From various corners one hears the sound of whooping cough. There are children suffering from jaundice and undoubtedly also from typhoid fever. . . . Their health deteriorates quickly, and there is no doubt that unless, at the very least, ameliorative measures are taken, within a month not 7 percent but 50 percent will perish. At the current moment, *not fewer than 80 percent* of children are sick.[20]

Medical supplies were never sufficient and usually only sent upon the outbreak of epidemics. Significant proportions of the labor force were chronically ill but still expected to drag themselves to the work site and keep up high output. Thousands of letters sent home to family members were intercepted by the police. Many tell harrowing stories of women giving birth in boxcars, children lost in the forest, workers crushed under the trunks of trees, children crying bitterly for lack of food, people hanging themselves or going crazy from hunger, drunk officials pilfering rations and abusing the settlers. "I would understand all of this if I were rich or if I had harmed someone," wrote one; "they drank our blood—that's all."[21]

At the Sixteenth Party Congress, which met June 26 to July 3, 1930, Stalin, in a stunning deviation from real events, asserted:

> Even the blind can see that the peasants have made a tremendous, a radical turn from the old to the new, from kulak bondage to free collective-farm life. There is no going back to the old. The kulaks are doomed and will be eliminated. Only one path remains, the collective-farm path. And the collective-farm path is no longer for us an unknown and unexplored path. It has been explored and tried in a thousand ways by the peasant masses themselves. It has been explored and appraised as a new path that leads the peasants to emancipation from kulak bondage, from want and ignorance. That is the basis of our achievements.[22]

No voice of dissent rang out. Yet the collective farms owed much to the dispossessed "kulaks" and little to the state: official statistics of the time indicated that 34.4 percent of the total capital of all collective farms had been seized from exiled fellow peasants.[23] Nor did most peasants eagerly join the movement: the secret police reported that countrywide, in April–July 1930, at least 240,000 peasants took part in 1,094 instances of rural unrest.[24]

Nor was there anything spontaneous about the formation of collective farms. On July 30, 1930, in districts of universal collectivization, the party-state abolished the peasant land society, the ancient institution that in the nineteenth century had inspired countless intellectuals to envision Russia passing directly to socialism without an exploitative and wrenching capitalist phase. None of that remained. Instead, collective farm chairmen and councils were appointed by Soviet officials. Typically, they had little knowledge of agriculture or rural affairs. That summer, 44.4 percent of the tens of thousands of militant workers sent to the countryside to spearhead collectivization (the "25,000ers") became chairmen of collective farms.[25] They joined various skilled employees—the equipment manager, the bookkeeper, the livestock manager, and the secretary—to form the collective farm "elite," typically a closely knit group of

Fig. 27 Artist not identified. "We will fulfill the grain harvesting fully and on time." 1929. "To the struggle against kulak sabotage of grain procurement."

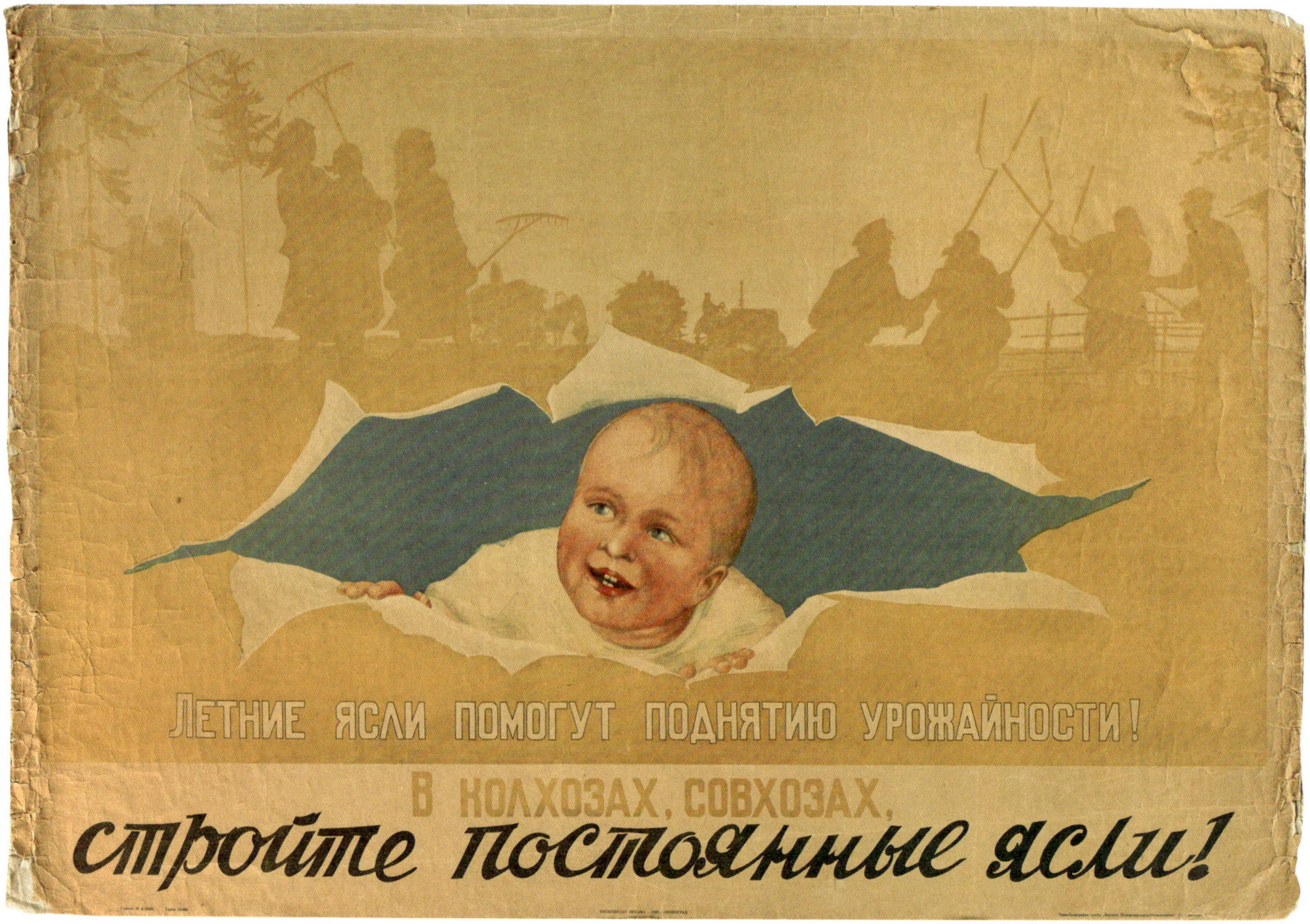

relatives and friends who appropriated for themselves the best resources of the farm. Similarly, the dense network of cooperatives that had formed during the NEP—and upon which Lenin, Bukharin, and other Bolshevik leaders had pinned such high hopes—was either crushed or nationalized.

The Politburo ramped up collectivization and the assault on "kulaks" throughout the fall. During the last four months of 1930, a further 1,120,000 households were pushed into collective farms.[26] Many senior secret police officials received the Order of the Red Banner for their work in implementing collectivization. Perhaps most astonishing, a central government decree on December 23 demanded the "dekulakization" of at least 3 percent of rural households. The fact that precise characteristics of "kulaks" were not supplied suggests that local officials were to unleash their ingenuity in ascribing to that category large numbers of rural folk. In fact, a pamphlet produced for the Leningrad region warned that any Soviet official who failed to ferret out "at least 5–7 percent" "kulaks" should be considered "a right opportunist."[27] There were regional variations. For example, in the Republic of Karelia an order of December 25 demanded that the "entire population take part in lumber production."[28] Under such circumstances (and those of the Komi Republic, in parts of which special settlers predominated), it must have been unclear who was a "kulak" and who was not.

Stalin and his supporters definitely triumphed on the "grain front" in 1930. The overall harvest was good, thanks largely

▲ **Fig. 28** Artist not identified. "Summer daycare helps raise crop yields! In collective farms, state farms, build permanent daycare!" 1929.

▶ **Fig. 29** V. N. Elkin. "Every collective farmer, every brigade, every MTS must know the plan for the Bolshevik sowing." 1931.

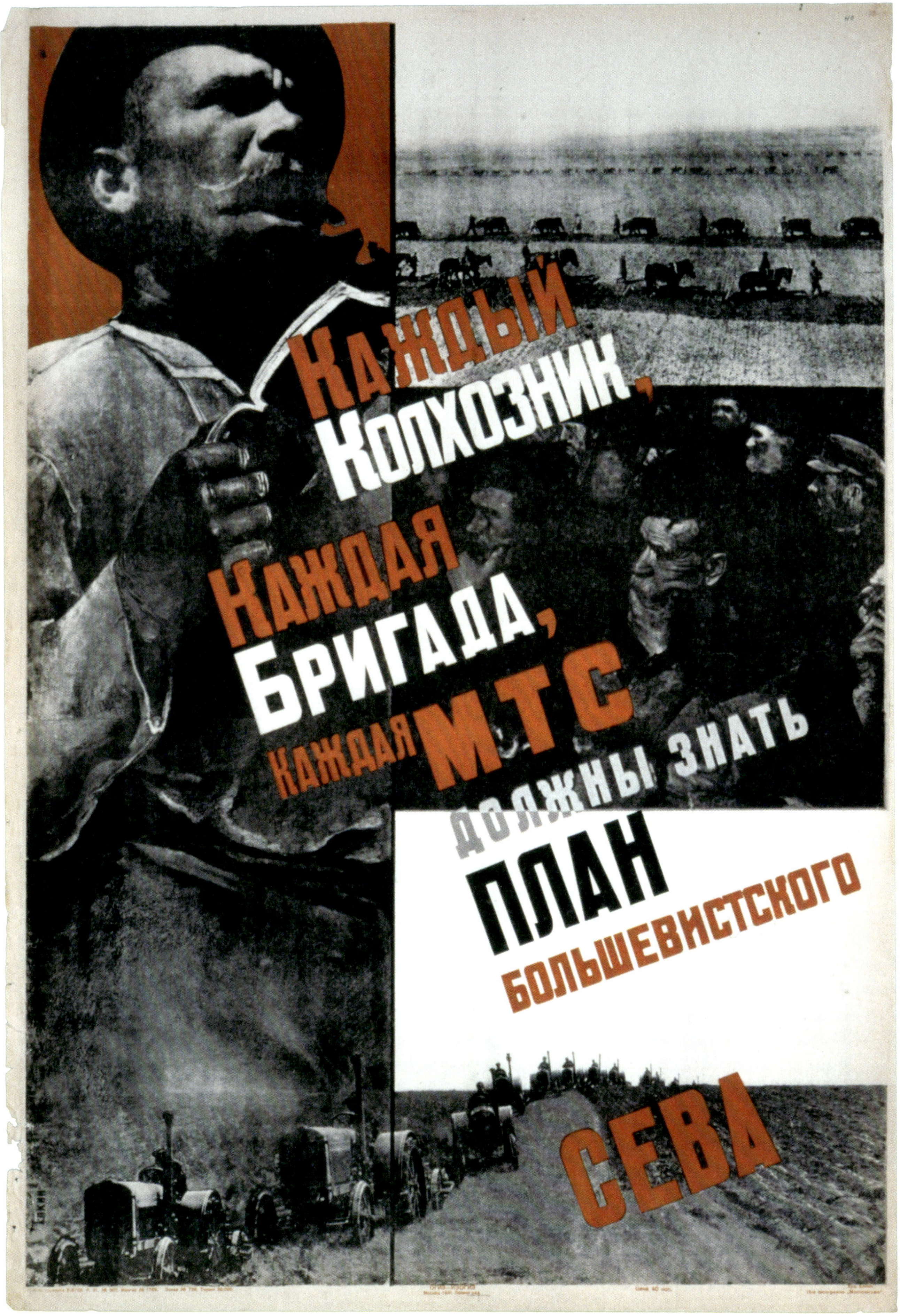
Каждый
Колхозник,
Каждая
Бригада,
Каждая МТС
должны знать
ПЛАН
большевистского
СЕВА
Цена 40 коп.

to favorable weather, the high number of individual households planting in fall 1929, and the mass exodus from collective farms in spring 1930. State grain collection was nearly twice that of 1928—between 35 and 40 percent of the entire harvest, compared to 20–25 percent in 1928. Such an astonishing success, from the point of view of economic planning and building socialism, had two main effects, aside from making possible dramatically bigger grain exports. First, it further enhanced the party leadership's already exaggerated confidence in its ability to "storm any fortress." Second, it gave the peasantry even less reason to work hard. Obviously, these two results were at odds with one another.

In fact, to keep the "successes" coming, in 1931 the leadership further broadened the definition of *kulak*. Henceforth official documents referred candidly to "economic and political kulaks," along with "unreliable elements,"[29] as all deserving despoilment and exile. It goes without saying that by this point only scattered remnants of "kulakdom" (*kulachestvo*), in the sense of well-to-do peasants, could have been found anywhere in the Soviet countryside. Nevertheless, several directives from February, both central and local, excoriated the "kulaks" and demanded their plunder and exile, for, as one decree intoned, "only the most merciless struggle against kulaks can ensure for us the victory in all spheres."[30] A Politburo directive of February 20 required the OGPU to prepare within six months one thousand more special settlements. Local officials hastened to fulfill these demands, subjecting at least two hundred thousand families (one million individuals) to "dekulakization" over the next several months.[31] Authorities at industrial operations around the country clamored for labor, while officials charged with herding peasants into collective farms were eager for a draconian means of both eliminating resistance and cowing others into submission. It was win-win for both parties.

Starting in March 1931, the party leadership devised ways to raise (or at least seem to raise) the standard of living of collective farmers. Their taxes were lowered, while those of "kulaks" and the remaining individual farmers, whose very existence was thought to provoke envy in collective farmers, were augmented. More funds were allocated to machine tractor stations (MTS) for farm machinery (the supply of tractors at MTS and collective farms increased from 520,000 to about 800,000 horsepower in January–June).[32] The "labor day" was introduced as a fancy scheme for increasing wages (it mostly did not).[33] The size of allowable individual garden plots was increased. Such policies were the carrot to the "dekulakization" stick.

"Kulaks" were supposed to be given carrots as well. A Politburo commission headed up by Andreev recommended in mid-May that they be supplied with basic provisions, tools, health care, and other benefits. The fact that the central authorities repeatedly criticized local officials for failing to maintain decent material conditions for the special settlers suggests that they in fact rarely did. With many tens of thousands of deported kulaks waiting in boxcars on train sidings in Magnitogorsk for deployment to factories and construction sites in May and June, it is hard to imagine officials tending carefully to the needs of such unfortunates. In the entire Ural region, the government planned to build 36,535 barracks, of which 21,131 were actually built by the end of 1931, leaving 32,300 families (over 150,000 people) in temporary lodgings.[34] Nor could they have coddled the 101,184 families (more than one-half million people) exiled to remote areas in the month of June alone.[35] The authorities built few laundries or disinfecting stations

Fig. 30 Artist not identified. "MTS–organizers of collective farm production." 1930. (In Ukrainian.)

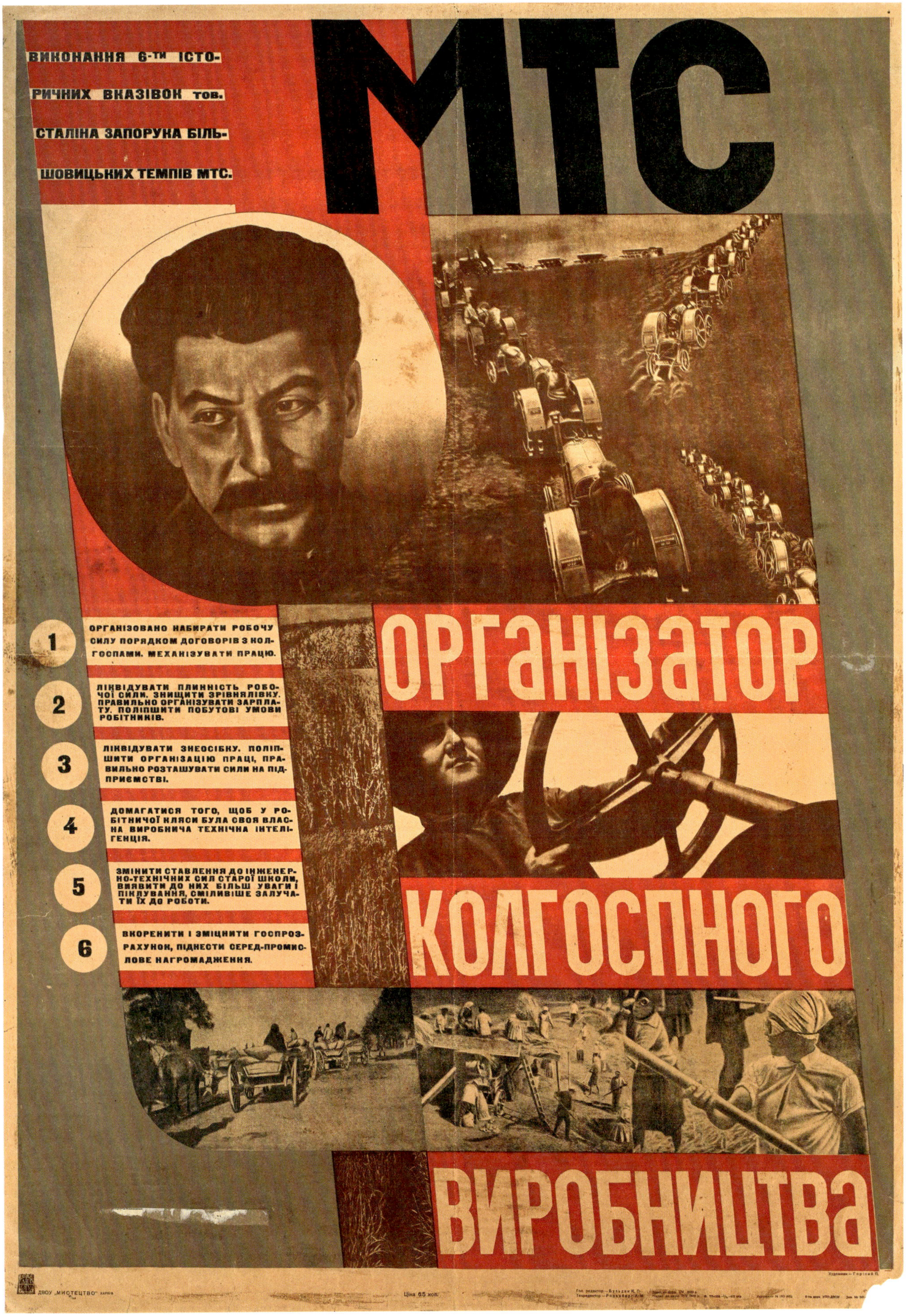
МТС
ВИКОНАННЯ 6-ТИ ІСТОРИЧНИХ ВКАЗІВОК тов. СТАЛІНА ЗАПОРУКА БІЛЬШОВИЦЬКИХ ТЕМПІВ МТС.
1 ОРГАНІЗОВАНО НАБИРАТИ РОБОЧУ СИЛУ ПОРЯДКОМ ДОГОВОРІВ З КОЛГОСПАМИ. МЕХАНІЗУВАТИ ПРАЦЮ.
2 ЛІКВІДУВАТИ ПЛИННІСТЬ РОБОЧОЇ СИЛИ. ЗНИЩИТИ ЗРІВНЯЛІВКУ. ПРАВИЛЬНО ОРГАНІЗУВАТИ ЗАРПЛАТУ. ПОЛІПШИТИ ПОБУТОВІ УМОВИ РОБІТНИКІВ.
3 ЛІКВІДУВАТИ ЗНЕОСІБКУ. ПОЛІПШИТИ ОРГАНІЗАЦІЮ ПРАЦІ, ПРАВИЛЬНО РОЗТАШУВАТИ СИЛИ НА ПІДПРИЄМСТВІ.
4 ДОМАГАТИСЯ ТОГО, ЩОБ У РОБІТНИЧОЇ КЛЯСИ БУЛА СВОЯ ВЛАСНА ВИРОБНИЧА ТЕХНІЧНА ІНТЕЛІГЕНЦІЯ.
5 ЗМІНИТИ СТАВЛЕННЯ ДО ІНЖЕНЕРНО-ТЕХНІЧНИХ СИЛ СТАРОЇ ШКОЛИ, ВИЯВИТИ ДО НИХ БІЛЬШ УВАГИ І ПІКЛУВАННЯ, СМІЛИВІШЕ ЗАЛУЧАТИ ЇХ ДО РОБОТИ.
6 ВКОРЕНИТИ І ЗМІЦНИТИ ГОСПРОЗРАХУНОК, ПІДНЕСТИ СЕРЕД-ПРОМИСЛОВЕ НАГРОМАДЖЕННЯ.
ОРГАНІЗАТОР
КОЛГОСПНОГО
ВИРОБНИЦТВА
ДВОУ „МИСТЕЦТВО" ХАРКІВ
Ціна 65 коп.

for the settlers. Nor was clean drinking water always available; in western Siberia, in November 1931, there was only one well for every nine hundred families (the minimum was supposed to be one well per hundred families).[36] Epidemics of typhus and other diseases inevitably broke out. Indeed, on October 26, 1931, Iagoda reported that in eastern Siberia, in the Narym region up the River Ob', children under three were dying at a rate of 8–12 percent per month, and in Magnitogorsk, at a rate of 15 percent per month, largely because of poor material conditions, including inadequate nutrition.[37] Similar disheartening reports flowed in to Moscow throughout the summer and fall. A temporary statute on the rights and obligations of special settlers, approved by Iagoda on October 25, allowed them to receive newspapers and books published in the USSR, to correspond and send and receive packages, to lodge complaints and petitions with local and central officials, and to arrange for children and elderly family members to live with relatives. Such promises may well have improved the quality of life of special settlers, though it is unclear to what extent they were fulfilled in practice.

Along with intensive "dekulakization," officials had achieved a high level of collectivization: 600,000 households in the first five months of 1931.[38] Yet discontent was clearly ongoing. Nearly 16 percent of all collective farms suffered "hostile disturbances" in the first six months of the year, many involving violent attacks on Communist activists, whose coercion and abuse that summer were described in a secret police report as "total lawlessness."[39] Not necessarily to rectify this situation, the Politburo decreed on July 20 that the rooting out of "kulaks" had been largely completed. Henceforth, "dekulakization" should occur solely on a case-by-case basis. It was also at this time that the special settlements were transferred to the jurisdiction of the OGPU and thus became a set of islands within the "gulag archipelago." Again, regional variations persisted; for example, officials in the Republic of Karelia launched another round on July 16, and local agents proceeded to target "economic and political kulaks" starting on July 23. The operation continued through September.

As with Lenin's "retreat on the economic front" of 1921, Stalin's halting of "dekulakization" involved battening down the achievements already obtained. Thus, on July 30 the central authorities decreed the socialization of livestock. Those who refused would have it taken by force, in exchange for the merest receipt. Massive flight from the collective farms, as well as the slaughter, sale, and abandonment of livestock, ensued. The leadership also wanted to win over the younger generation and turn them away from their parents by housing them in barracks with teachers, clubs, rural correspondents, and Komsomol activists inculcating Soviet values. Of course, in the early 1930s available resources made it difficult to achieve this goal.

The party leadership was consumed for the rest of the year with collecting and exporting grain. Stalin and his closest supporters kept up steady pressure on the grain-producing regions. The regional party bosses, for their part, under heavy resistance from the local populations, repeatedly pleaded for a lowering of their targets. Sometimes Stalin gave in, though it is not always clear why. Moreover, he and his henchmen often viciously dressed down regional party leaders for seeking delivery reductions. At the fall plenum, on October 31, Mikoian rebuked Filipp Goloshchekin (1876–1941), the party boss of Kazakhstan: "Comrade Goloshchekin, I have read out to you an official document, a decision of the Politburo: 55 million [poods, equal to 900,000 tons] without

Fig. 31 V. Antipenko. "Rural correspondent! Join the ranks of the shock worker brigades for socialist agriculture." 1931.

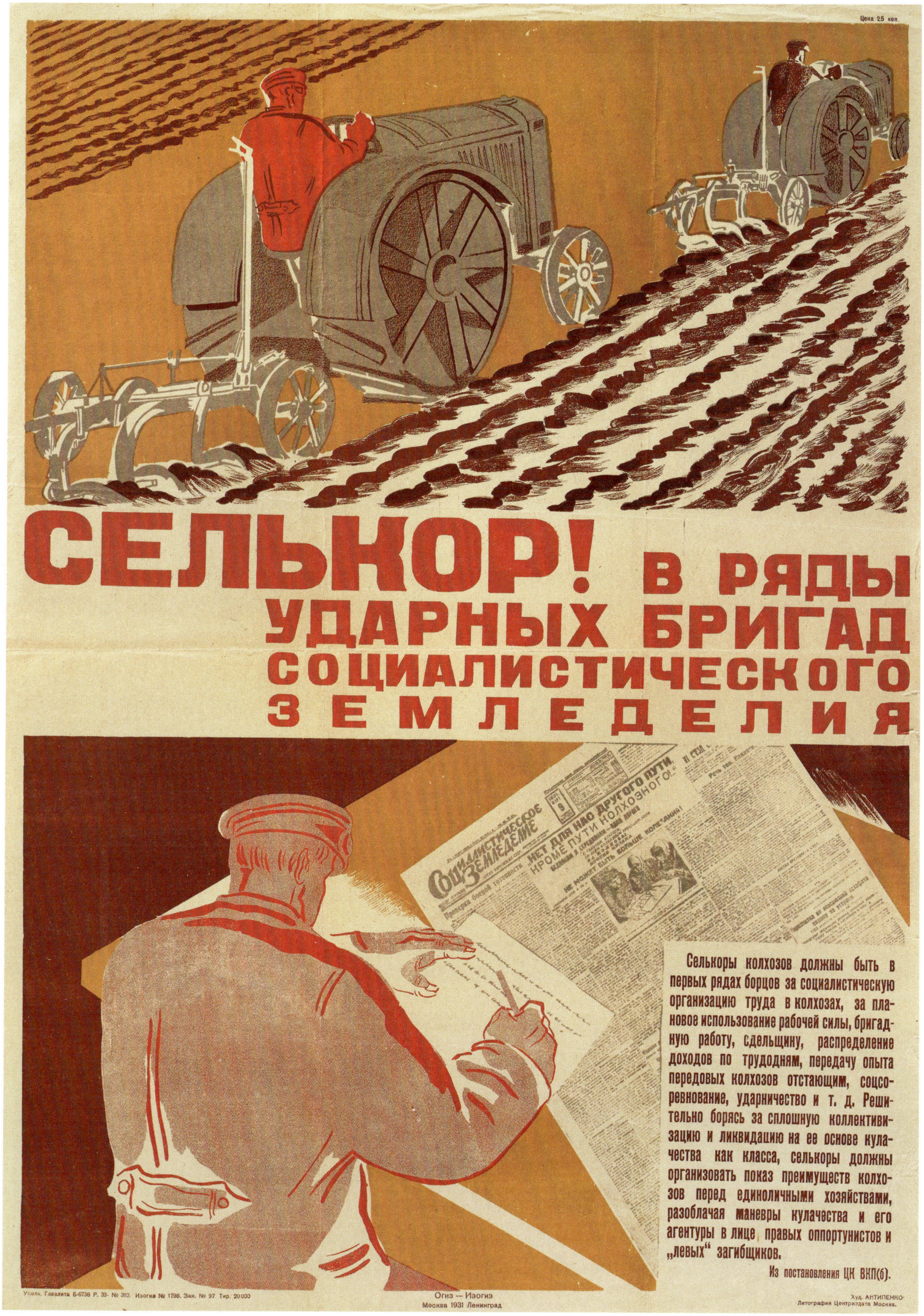

Цена 25 коп.
СЕЛЬКОР! В РЯДЫ УДАРНЫХ БРИГАД СОЦИАЛИСТИЧЕСКОГО ЗЕМЛЕДЕЛИЯ
Социалистическое земледелие
НЕТ ДЛЯ НАС ДРУГОГО ПУТИ, КРОМЕ ПУТИ КОЛХОЗНОГО!
Селькоры колхозов должны быть в первых рядах борцов за социалистическую организацию труда в колхозах, за плановое использование рабочей силы, бригадную работу, сдельщину, распределение доходов по трудодням, передачу опыта передовых колхозов отстающим, соцсоревнование, ударничество и т. д. Решительно борясь за сплошную коллективизацию и ликвидацию на ее основе кулачества как класса, селькоры должны организовать показ преимуществ колхозов перед единоличными хозяйствами, разоблачая маневры кулачества и его агентуры в лице правых оппортунистов и „левых“ загибщиков.
Из постановления ЦК ВКП(б).
Уполн. Главлита Б-6738 Р. 33· № 310. Изогиз № 1798. Зак. № 97 Тир. 20000
Огиз — Изогиз
Москва 1931 Ленинград
Худ. АНТИПЕНКО
Литография Центриздата Москва.

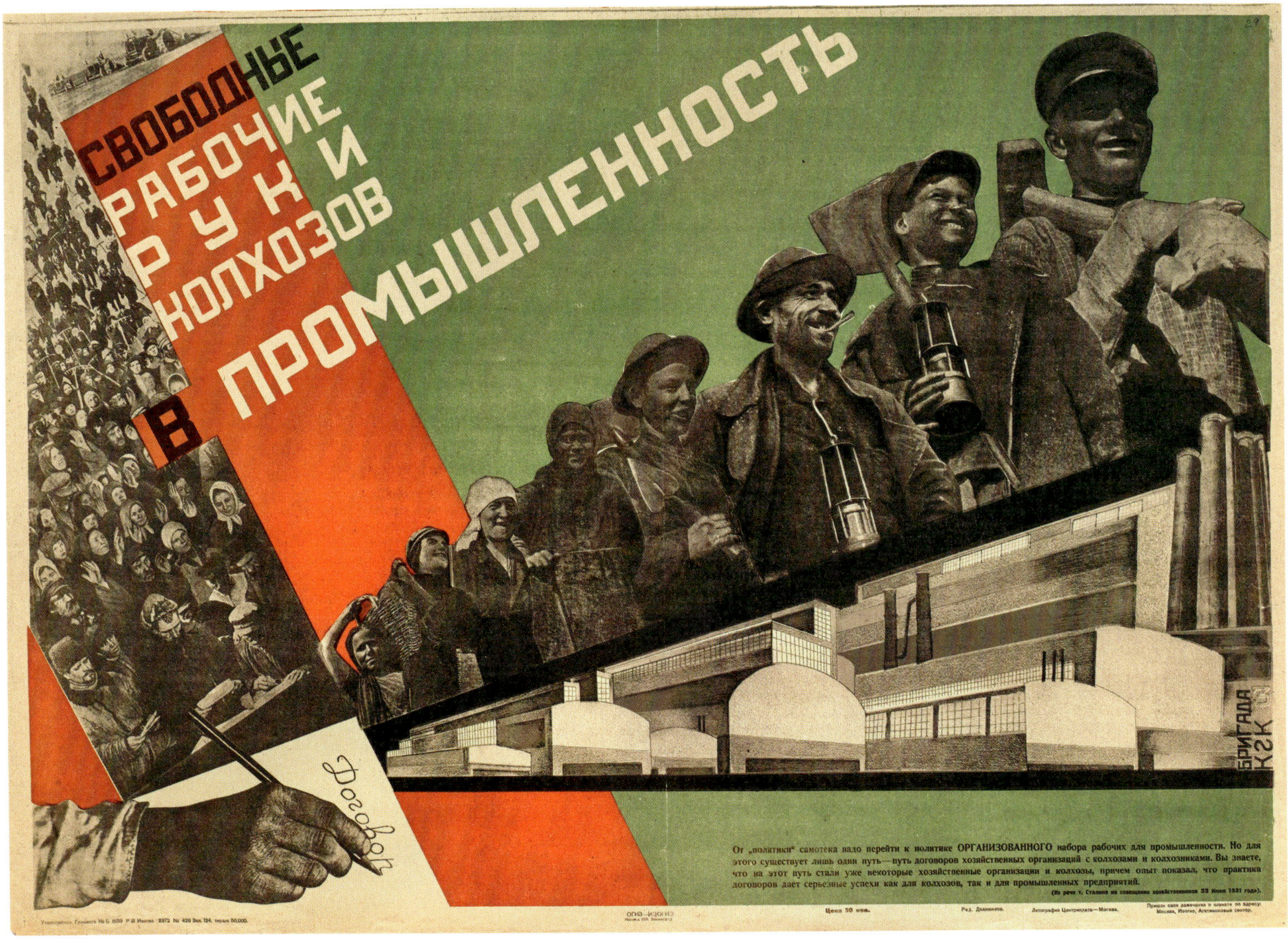

Fig. 32 V. Koretsky. "Free working hands of the collective farms—to industry." 1931.

rice. This is absolutely precise, and I don't know why you are confusing things."[40] In a telegram of December 5, Stalin and Molotov warned local officials to use the harshest measures to ensure proper grain collection. The head of a district party committee in the Middle Volga region, Tkachev, had the telegram printed in a district newspaper, presumably hoping to show that he was not personally responsible for the harsh measures he had to enforce. He lost his party card. In turn, the party bosses held their subordinates' feet to the fire. Hundreds of local officials who failed to keep up the pressure were dismissed. Thousands were prosecuted in the course of the collection campaign that fall.

Throughout the autumn, complaints from rural dwellers flooded into newspapers, central party and government institutions, and the chancelleries of top officials, like Stalin and Kalinin. Such grievances seeped out to the West. On October 12, the London *Times* quoted a peasant lamenting: "It is terrible here in the Kolkhoz. We cannot speak or we shall be sent away to Siberia as they sent the others. We are afraid. I had three cows. They took them away. . . . We dare not oppose the Communists or we shall be exiled. We have to keep quiet."[41] And little wonder: inhabitants of the Lower Volga region had to make do with an average of 101 kilograms of grain yearly, as compared to an average in the 1920s of 210 kilograms.[42] The number of people receiving regular rations—nearly all of them in urban areas—rose from 26 to 33 million.[43] (A smaller, more select group on the Special List and List 1 received much bigger rations.) Little wonder that millions of peasants fled from the collective farms into the city. A monthly breakdown is unavailable, but roughly 4 million peasants—not counting exiles—migrated to urban areas in 1931.[44]

Despite the leaders' success in pushing the peasantry into collective farms, the balance sheet was bleak. In 1930–31, at least 381,026 families—1,803,392 individuals—were exiled as "kulaks."[45] Those figures exclude those who fled or died on the way to their special settlements. Available livestock sank dramatically in 1931: cattle by 7.2 million head, sheep and goats by 15.6 million, swine by 2.8 million, and horses by 6.6 million.[46] The grain harvest stood at roughly 538 million centners in 1931, down from 804 million in 1930. Collections, by contrast, increased, as did exports—51.8 million centners in 1931, up from 48.4 million in 1930. Worse still, the crop yield throughout the USSR tumbled from 48.7 poods/hectare in 1929 to 38.4 in 1931.[47] The primary reason for the decline seems to have been a falling work ethic. Since people were not being paid adequately for their work, they got into the habit of not working adequately. They planted less, failed to weed their fields, harvested without enthusiasm, and delivered crops negligently. All such facets of the collective farm system were bound to spell trouble. Ongoing forced collections, coupled with some unfavorable weather, meant disaster in 1932.

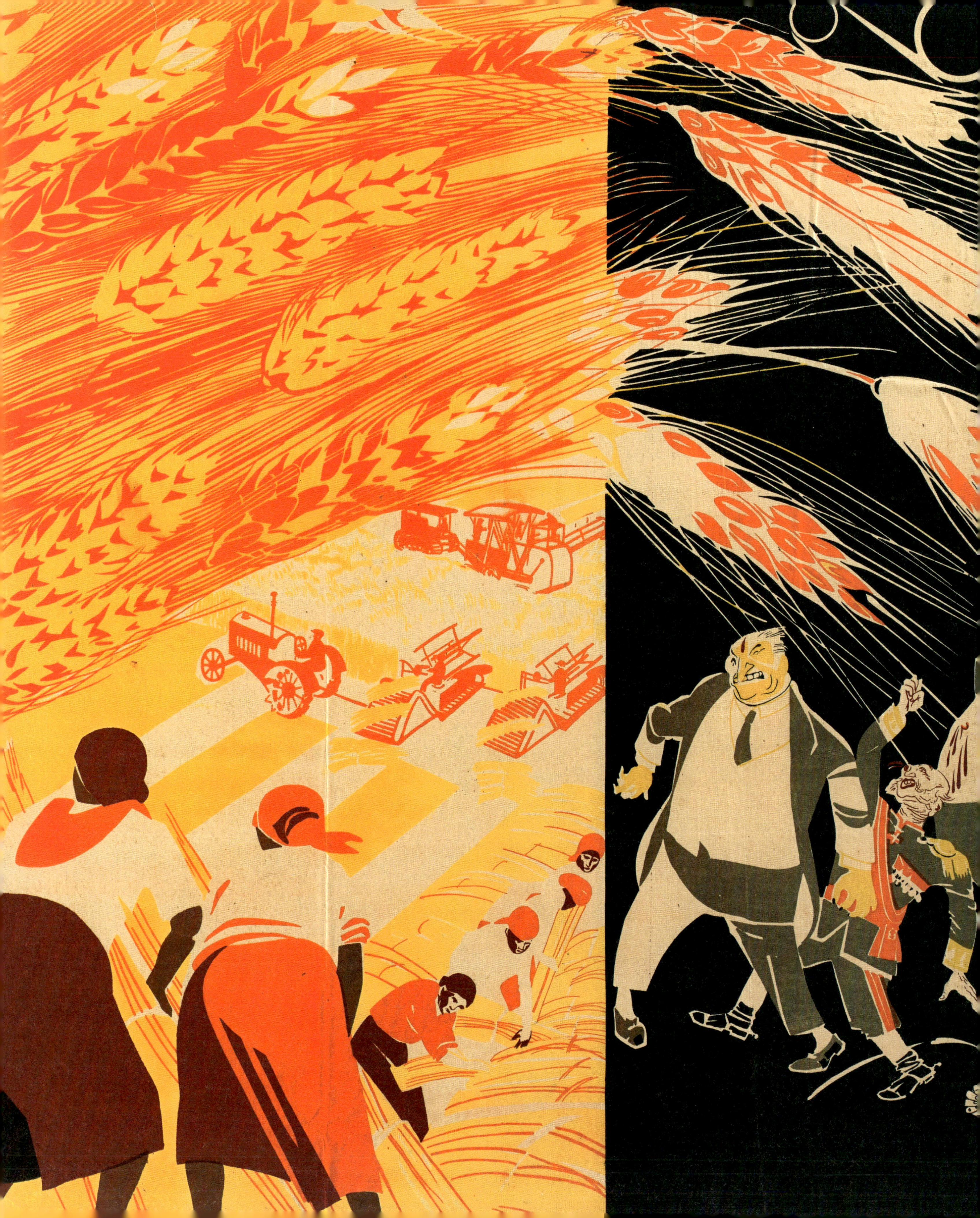

Chapter 3

DEMOGRAPHIC CATASTROPHE, 1932–1933

This chapter is the most tragic and heart-wrenching, but also the one for which almost no evidence is to be found in Soviet posters of the time. And no wonder—the government completely denied the famine's existence both while it was happening and for many years thereafter. Nevertheless, one poster, depicting an armed guard "vigilantly protecting the socialist harvest" (figure 36), correlates with the draconian law of August 7, 1932, which prescribed up to ten years' incarceration for gleaning even small quantities of grain from the fields. Representations in posters of a joyful life of abundance on collective farms, produced at the very time when famine was raging in the countryside, provide an antithesis to the true story, as a rising tide of baleful reports throughout 1932 warned central officials of widespread starvation and even cannibalism. Despite these reports, forced collectivization continued, food shipments to hungry regions were impeded, domestic passports were imposed in December 1932 to prevent the free movement of peasants, roadblocks also kept the starving from migrating to areas of relative plenty, and ultimately some 5–7 million people perished, mostly in the late spring of 1933.

In the winter of 1931–32, "storm brigades" were sent house to house, sometimes with armed police, "day and night" to conduct searches for grain, "not distinguishing between the poor, middle, and well off."[1] Central government orders of February 7 and 22, 1932, demanded that collective farmers deliver most of their seeds intended for planting. Many farmers refused to hand them over without a struggle. The government lashed out against "speculators." On March 10, the London *Times* reported that the OGPU was arresting "individual non-proletarians, especially former NEPmen, and citizens whose source of livelihood is not apparent, and demanding an arbitrarily fixed tax or ransom, mostly expressed in gold or foreign currency, before liberating them. Persons refusing or unable to pay the ransom remain in prisons and labour camps indefinitely, charged with such obscure crimes as speculation or attempting to undermine the financial stability of the State."[2] In reality, very little wealth remained in private hands in the countryside. One eyewitness visited a village with one thousand households; he saw "only two chickens, a rooster, and occasionally a mangy dog."[3] Another, the novelist and Nobel Prize winner Mikhail Sholokhov (1905–84), wrote to Stalin on April 20 that in one district of the North Caucasus region, as of February 1, there were only eighteen households (out of 13,629) with two cows.

Politburo decrees of January 28 and February 3, 1932, chided local officials for failing to improve living conditions and services provided to special settlers, such as health care, sanitation, cultural offerings, and even mail delivery. One report indicated that as of January, only 40–49 percent of special settler children were attending elementary school in the Ural region, western Siberia,

◀ Detail of "Grain is our strength," *see page 80*.

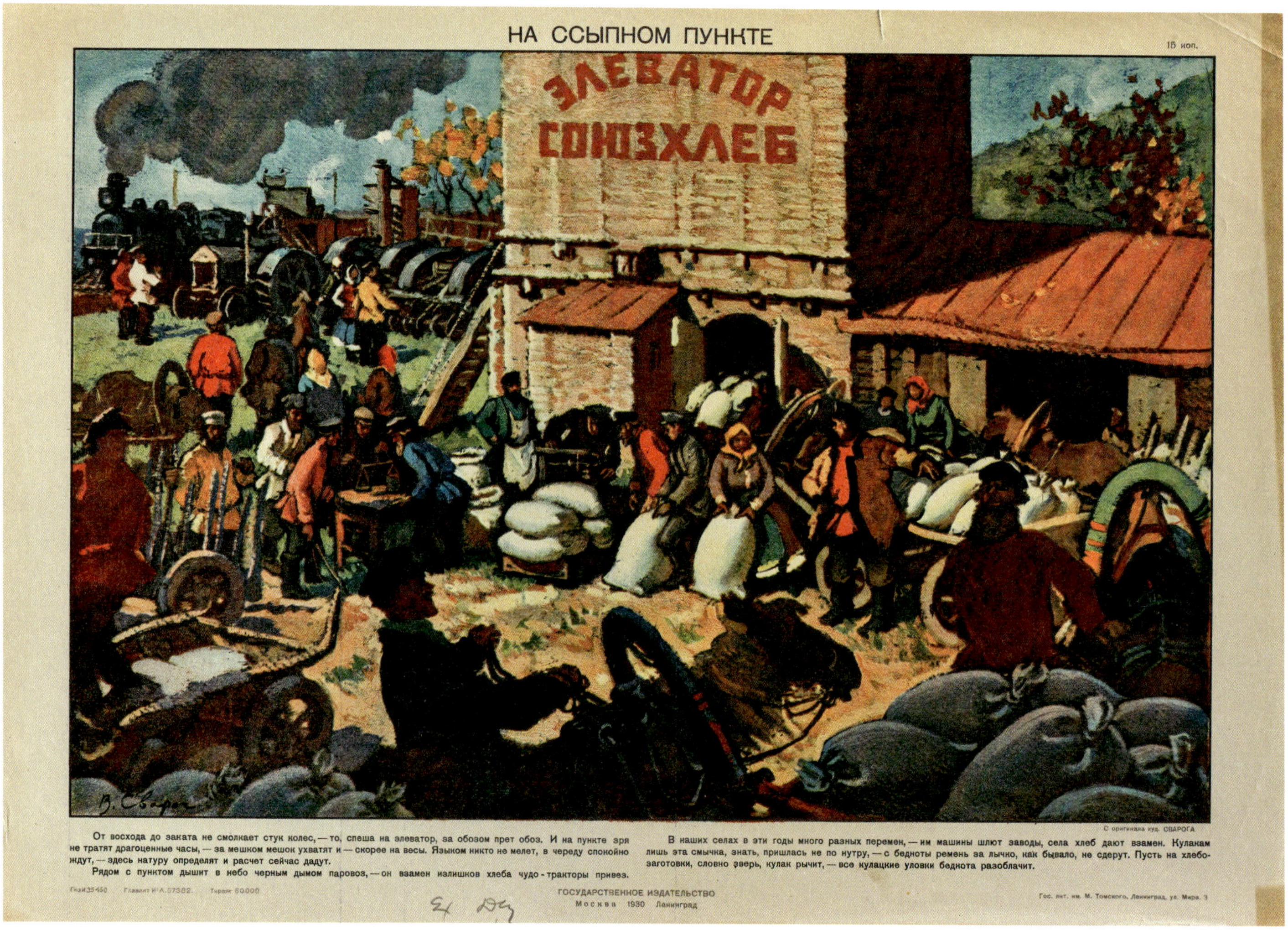

▲ **Fig. 33** V.S. Svarog (Korochkin). "At the grain delivery station." 1930. ". . . In our villages in these years there have been many changes: factories send them machines, and they give grain in exchange . . ."

▶ **Fig. 34** N. Kochergin. "With militant tempos let us fulfill the decree of the government on the agricultural tax for 1932." 1932. "The full and timely payment of the agricultural tax is a guarantee of victory in the struggle for organizational and economic strengthening of the collective farms and social and cultural reorganization of the countryside."

and the Northern Territory.[4] Such schools as they attended, moreover, were badly supplied with materials, and in many areas most of the teachers were themselves special settlers. For a state committed to agitation and propaganda and reeducating the children of "kulaks," this was a fiasco.

Reports at the local level were even bleaker. At the Shkal'nye mines in the Republic of Karelia, the housing was "crowded, cold, damp." There was inadequate firewood and no children's food at all. The number of milk cows had dropped from sixty-seven to twenty-eight. As for the "red circle," "no one comes because it is cold, damp, and filthy."[5] Of course, there was no hope of improving conditions in the special settlements without massive state investments, which party leaders had no intention of providing. The same went for a Central Committee decree of March 26, obligating the party "to make sure that every collective farmer has his own cow, small livestock, and fowl."[6] Yet where were these animals to come from, given that tens of millions had been slaughtered? The same went for orders of May 6 and 10, which authorized collective farmers to sell grain and meat on the newly legalized "collective farm markets" after the plan had been fulfilled. In reality, the collections were so steep that nothing was ever left to sell;

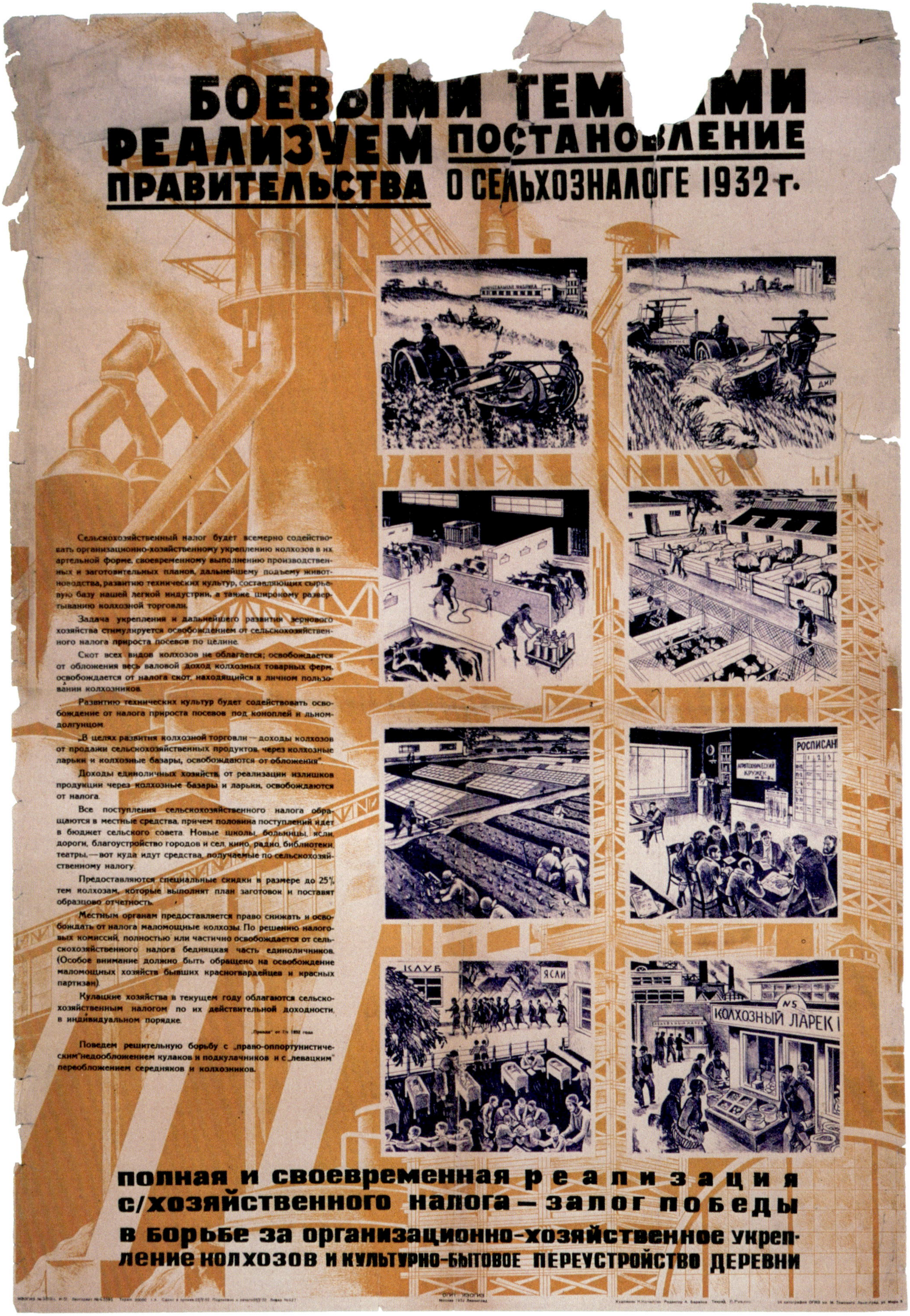
БОЕВЫМИ ТЕМ МИ
РЕАЛИЗУЕМ ПОСТАНОВЛЕНИЕ
ПРАВИТЕЛЬСТВА О СЕЛЬХОЗНАЛОГЕ 1932 г.
Сельскохозяйственный налог будет всемерно содействовать организационно-хозяйственному укреплению колхозов в их артельной форме, своевременному выполнению производственных и заготовительных планов, дальнейшему подъему животноводства, развитию технических культур, составляющих сырьевую базу нашей легкой индустрии, а также широкому развертыванию колхозной торговли.
Задача укрепления и дальнейшего развития зернового хозяйства стимулируется освобождением от сельскохозяйственного налога прироста посевов по целине.
Скот всех видов колхозов не облагается; освобождается от обложения весь валовой доход колхозных товарных ферм, освобождается от налога скот, находящийся в личном пользовании колхозников.
Развитию технических культур будет содействовать освобождение от налога прироста посевов под коноплей и льном-долгунцом.
„В целях развития колхозной торговли — доходы колхозов от продажи сельскохозяйственных продуктов через колхозные ларьки и колхозные базары, освобождаются от обложения".
Доходы единоличных хозяйств от реализации излишков продукции через колхозные базары и ларьки, освобождаются от налога.
Все поступления сельскохозяйственного налога обращаются в местные средства, причем половина поступлений идет в бюджет сельского совета. Новые школы, больницы, ясли, дороги, благоустройство городов и сел, кино, радио, библиотеки, театры, — вот куда идут средства, получаемые по сельскохозяйственному налогу.
Предоставляются специальные скидки в размере до 25% тем колхозам, которые выполнят план заготовок и поставят образцово отчетность.
Местным органам предоставляется право снижать и освобождать от налога маломощные колхозы. По решению налоговых комиссий, полностью или частично освобождается от сельскохозяйственного налога бедняцкая часть единоличников. (Особое внимание должно быть обращено на освобождение маломощных хозяйств бывших красногвардейцев и красных партизан).
Кулацкие хозяйства в текущем году облагаются сельскохозяйственным налогом по их действительной доходности, в индивидуальном порядке.
Поведем решительную борьбу с „право-оппортунистическим" недообложением кулаков и подкулачников и с „левацким" перобложением середняков и колхозников.
КЛУБ
ЯСЛИ
№5
КОЛХОЗНЫЙ ЛАРЕК
полная и своевременная реализация с/хозяйственного налога — залог победы в борьбе за организационно-хозяйственное укрепление колхозов и культурно-бытовое переустройство деревни

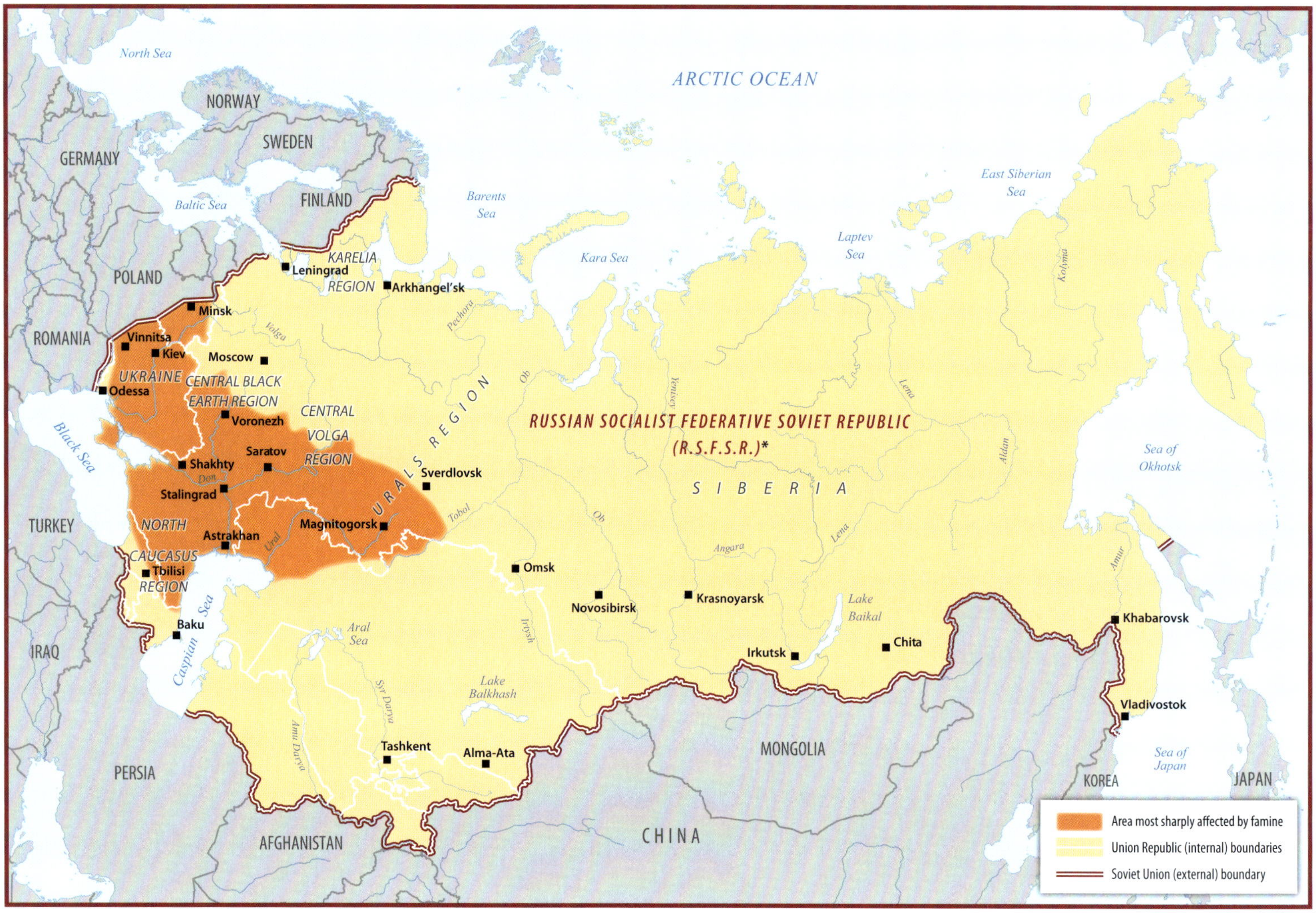

*Country name changed to Russian Soviet Federative Socialist Republic in 1937.

Map 3 Areas most sharply affected by famine, 1932–1933

there was barely enough to eat and to set seeds aside for planting. Indeed, peasants were to such an extent driven to steal grain just to survive that on April 13 the Politburo called for putting on five to ten show trials against "organizers" of grain theft as enemies of the people.[7]

It was perhaps not entirely a coincidence that precisely at this time, on April 23, 1932, the party issued the decree "On the Reorganization of Literary and Artistic Organizations," which established socialist realism as the official and only permissible mode of artistic expression in the USSR. (Stalin was reportedly the leading advocate of the decree.) By adopting the "correct" party, ideological, class, and popular point of view, the artist would naturally depict the building of socialism, Soviet society as it would become and not as it was; the bright, promised future, not the tawdry, depressing present.

The present, sadly, was growing more and more depressing. On March 23, 1932, the Politburo reduced the grain allocation for the 20 million people on ration Lists 2 and 3.[8] The sowing campaign went badly, despite promises of prizes in money and in kind for fulfilling the plan: the farmers who went out into the fields were often hungry and weak. Many of those with energy did not work their hardest; what was the point? It certainly did not help that the total draft power (horses, oxen, and tractors) in the USSR had plummeted from 30 million in 1928 to 23 million in 1932—a serious impediment to timely fieldwork.[9] The collectivization drive also stalled. In the Russian Republic, 1,370,800 households abandoned the collective farms, as did 41,200 in Ukraine, in the first half of 1932.[10] Even the press, which was usually so reticent to admit failings, reported on "temporary difficulties" of the collective enterprise, though "rich peasants" engaging in "sabotage" were typically presented as the culprits.[11] Simon Starow (1916–98), writing under the name Miron Dolot, recalled that in late May a new set of urban activists descended upon his village a hundred miles south of Kiev, the Ukrainian capital, to make preparations for the harvest. "In no time at all," he wrote, they "took over our entire village like tyrants, imposing their wills and demands upon us."[12]

Throughout the summer, Stalin and other party leaders received alarming reports from the countryside. One report, of June 18, 1932, warned that tens of thousands of Ukrainians were wandering away from their villages seeking food.[13] Stalin was furious and demanded strict measures to prevent such movement. He seems to have discounted reports and rumors of starvation and even cannibalism in Ukraine, as exaggerations and sneaky peasant ploys to avoid handing over surplus grain. He deeply mistrusted the peasantry, feared their wily tricks. It did not help that he received conflicting reports from top officials. Earlier in the month, Stanislav Kosior (1889–1939), the party boss in Ukraine, had written to Stalin, admitting only to some isolated instances of hunger. Yet the chairman of the Central Executive Committee of Ukraine, Grigory Petrovsky (1878–1958), had just written (on June 10), warning of dire famine conditions in many places.[14] When Stalin received word, on July 17, of mass departures from collective farms, he demanded harsh repression against instigators. Fearing a loss of the harvest, in July Stalin sent his toughest operators, Kaganovich and Molotov, to Ukraine to ensure fulfillment of collections. Although they furnished little or no food aid, they did work to bring manufactured goods to the countryside, as an incentive for peasants to hand over their grain.

But Stalin was ready to play hardball. In a letter to Molotov on July 20, 1932, he decried the "lenient" punishments for the theft of state property from collective farms and other institutions—a practice

Цена 50 коп.

С БОЛЬШЕВИСТСКОЙ ЭНЕРГИЕЙ ОСУЩЕСТВИМ ПЛАН ПОСЕВА 43 МЛН. ГА ОЗИМЫХ КУЛЬТУР

На основе социалистического соревнования и ударничества, мобилизуя внимание общественности вокруг осенней посевной кампании, выполним план:

1. По ремонту машин и с.-х. инвентаря, который должен быть закончен к 1—15 августа.

2. По сбору колхозного фонда зерновых семян озимых культур, приступив немедленно после уборки и обмолота урожая, создавая его из лучшего по качеству семенного зерна.

Доведем за 10 дней до начала работ до каждой бригады, рабочего и колхозника оперативное задание по всем работам осеннего сева.

Обеспечим полный перевод всех работ по осенней кампании на сдельную оплату.

Отсортируем семенной материал и по указанию агронома произведем протравливание зерна против вредителей.

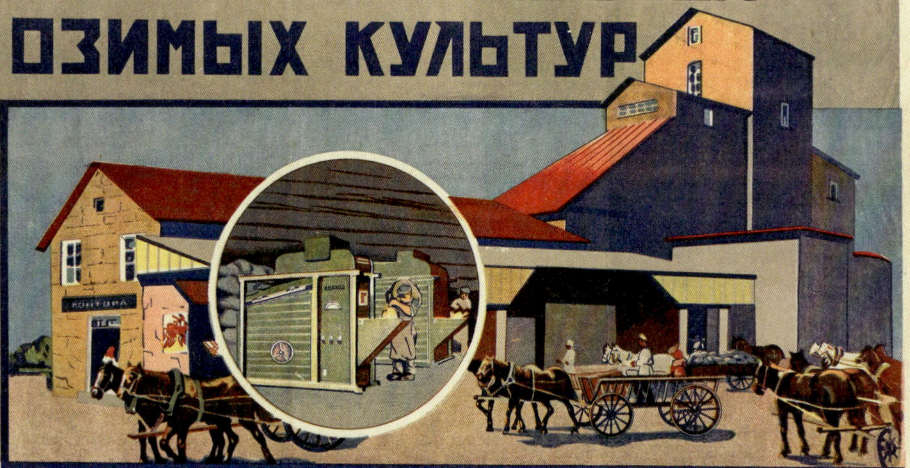

ПРАВИЛЬНАЯ ОРГАНИЗАЦИЯ ТРУДА—МОГУЧИЙ РЫЧАГ КОЛЛЕКТИВИЗАЦИИ

Проверка высева.

Перед посевом следует произвести проверку сеялки на высев зерна. Для этого сеялку закатывают на полотнище брезента, наполняют семенной ящик зерном и приподнимают раму сеялки так, чтобы колесо с шестерней вращалось в воздухе. На колесе делают метку мелом, чтобы можно было отсчитать число оборотов.

Затем измеряются обод колеса и захват сеялки. Предположим, обод колеса=4 метрам и захват=4 метрам. Умножаем 4 на 4, получаем 16 кв. метров—площадь, которую засевает сеялка при одном обороте колеса.

Всего в одном гектаре—10 тыс. кв. метров. Разделим 10 тыс. на 16 и получим 625. Значит, колесо сеялки должно обернуться 625 раз, чтобы засеять гектар, или 10 тыс. кв. метров.

Для засева одной десятой части гектара колесо должно обернуться 62 раза. Обернем наше колесо 62 раза. Высыпавшиеся на брезент семена соберем и взвесим. Теперь мы точно знаем, сколько высевает наша сеялка на одну десятую гектара.

Предположим, что при 62 оборотах высыпалось семян 12 килограммов. Это—на десятую гектара. На полный гектар это составит 120 килограммов. Если надо изменить количество высева, то передвигают рычажок на доске. Потом снова проверяют, вращая колесо.

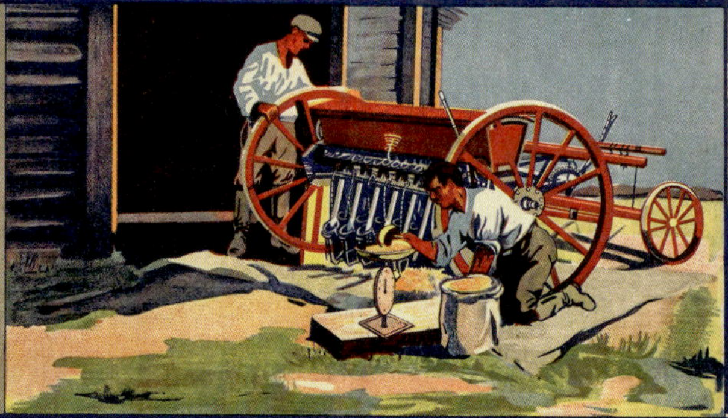

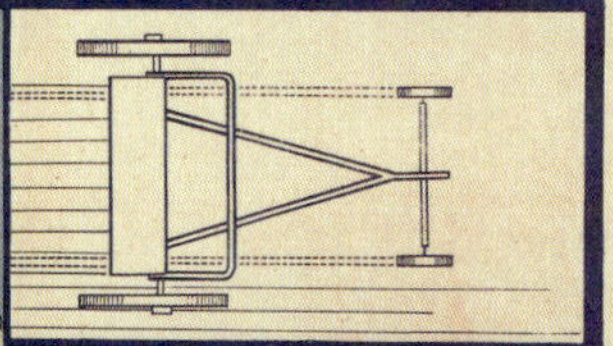

Перед посевом следует произвести установку передка сеялки. Если мы имеем 17-рядовую сеялку с шириной междурядий в 12 сантиметров, то рабочая ширина захвата сеялки будет—17×12=204 сантиметрам. Ширина всей сеялки—216 сантиметров Рабочую ширину захвата сеялки помножить на 2. Из полученного вычесть ширину сеялки. Полученная цифра будет расстоянием между колесами передка. При таком подсчете у 17-рядовой сеялки расстояние между колесами передка будет 192 сантиметра.

Производственная бригада—основа коллективного хозяйства

ПОМНИ ГЛАВНЫЕ УСЛОВИЯ ПРАВИЛЬНОГО ПОСЕВА.

1. Семена должны располагаться на равном расстоянии друг от друга.
2. Они должны попадать в сырую землю.
3. Они должны быть заделаны на одинаковую глубину.

При посеве рядовой сеялкой расходуется меньше семян, чем при разбросном посеве. Рядовой сеялкой сберегается от 15 до 50 килограммов зерна на гектар. Урожай увеличивается на 10—20%.

Каждая тысяча гектаров, засеянная рядовой сеялкой, дает экономии 30—40 тонн зерна и увеличивается урожай не меньше, чем на 150 тонн.

Надо полностью использовать производительность сеялки. При посеве следует применять двух- или трехсменную работу; тогда одна сеялка за сезон может засеять до 160 гектаров.

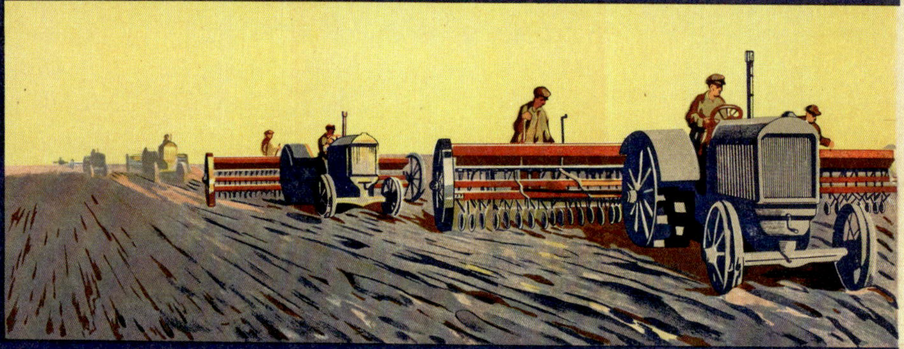

ТРУДОДЕНЬ—ОСНОВА УЧЕТА ТРУДА В КОЛХОЗЕ

Чтобы избежать грубых ошибок при посеве, надо, хотя бы раз в день, проверять работу высевающего аппарата сеялки на ходу.

Для этого до начала работы вдоль внутренних стенок семенного ящика, приблизительно на треть его высоты, краской по линейке проводят черту. Нижнюю часть ящика засыпают зерном, разравнивая его дощечкой так, чтобы заполнить его ровно до черты. После этого засыпают в ящик столько зерна, сколько надо на посев гектара; затем отмеряют в поле один или несколько гонов, по площади равных гектару, и после прохода этих гонов проверяют, понизился ли уровень зерна в семенном ящике до черты.

Если окажется, что этот уровень выше или ниже нашей отметки, то надо сейчас же соответствующим образом изменить положение указателя рычага вала катушки.

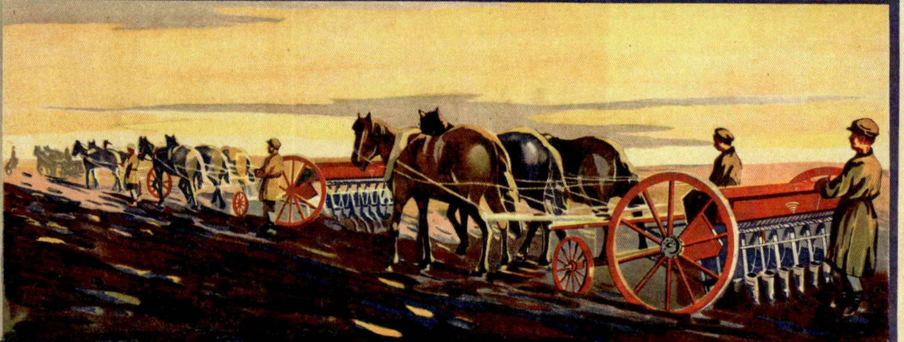

Путем соцсоревнования бригады должны дать наивысшую производительность сеялок в осенний сев

Ук. Главлита № 8040. Р. 31. № 374. Изогиз № 2218. Заказ № 471. Тираж 20 000.

ОГИЗ—ИЗОГИЗ
Москва 1931 Ленинград

С оригинала худ. А. В. СОКОЛОВА. Автор А. И. РЕМОРОВ
Редактор А. ЛАДОХИНА
5-я лит. „Мособлполиграф"

he attributed to anti-Soviet elements.[15] Instead of six or eight months of actual jail time for such crimes, he proposed from ten years in prison to capital punishment. On August 7, the Politburo issued precisely such a degree. Stalin also demanded a press barrage praising the law. On August 21, for example, the agrarian economist Iakov Nikulikhin (1896–1939) called it a "powerful lever for educating the collective farm masses about socialist relations to socialized means of production . . . a powerful tool in the struggle against the opposition of the kulaks." He added that this law had been "greeted with euphoria by all workers and all conscious collective farmers."[16] Of course, nothing could be further from the truth.

The results of the legislation were horrific. As of January 1, 1933, in the Russian Republic alone 54,645 people were sentenced and 2,100 executed.[17] One woman, a twenty-eight-year-old with three children, received a ten-year sentence for taking a total of 4 kilograms of grain. An entire family was sentenced for catching fish in a river flowing past their collective farm. A sixty-five-year-old was condemned to ten years for gathering twelve stalks of grain. Such instances could be multiplied. Firewood and lumber in the forest were also protected. One peasant lamented, "Just to get warm, you have to take risks every single day, wear yourself out."[18] The London *Times* reported on August 15 that to prevent the loss of grain during rail shipment, the government was forming squads of Communist youth charged with "not only the inspection of grain defenses, but the incitement of harvesters to greater efforts."[19] Fear of arsonists setting fire to fields of grain was, according to the newspaper, the pretext for prohibiting peasants from carrying firearms or hunting in the central regions of the Russian Republic. This was presumably merely a pretext for disarming ordinary people.

Grim conditions worsened. The peasantry in grain-producing regions, especially Ukraine and the North Caucasus, was growing extremely restive. Mass disturbances broke out dozens of times in August and September. The availability of food even in urban areas deteriorated. In July–September, some 11 million workers received monthly meat rations of up to 1 kilogram (for their entire family)—compared to 40 million with the right to bread rations. Only the 380,000 most privileged white-collar workers received a meat ration.[20] Butter and milk were even harder to get. Wheat rust and other diseases struck the crops. Weeds choked them. Harvest targets were revised downward to 67–71 million tons in late August and, in secret reports, to 65 or even 60 million tons (the actual harvest may, in fact, have been as low as 55 million).

Panicked, the Politburo set up "extraordinary commissions" for grain collection in Ukraine and the North Caucasus on October 22, 1932, headed by Molotov and Kaganovich, respectively. Kaganovich was merciless. On November 4, he issued an order "to crush sabotage against collection and planting, which is being organized by kulak counterrevolutionary elements, to destroy the resistance of rural communists, who de facto are acting as perpetrators of sabotage, and to liquidate the un-communist passivity and willingness to compromise with saboteurs." To scare the countryside, he demanded the blacklisting of three whole towns: to deny them all services, all exchanges, all trade, all support; to purge collective, cooperative, and state institutions of "all manner of alien and hostile elements"; and to threaten worse reprisals for "further sabotage."[21] Throughout the North Caucasus some fifteen thousand people were arrested.[22]

In early November, Stalin and Molotov cabled party bosses, threatening them

Fig. 35 A.V. Sokolov. "With Bolshevik energy, we will fulfill the plan for 43 million hectares of winter sowing." 1931.

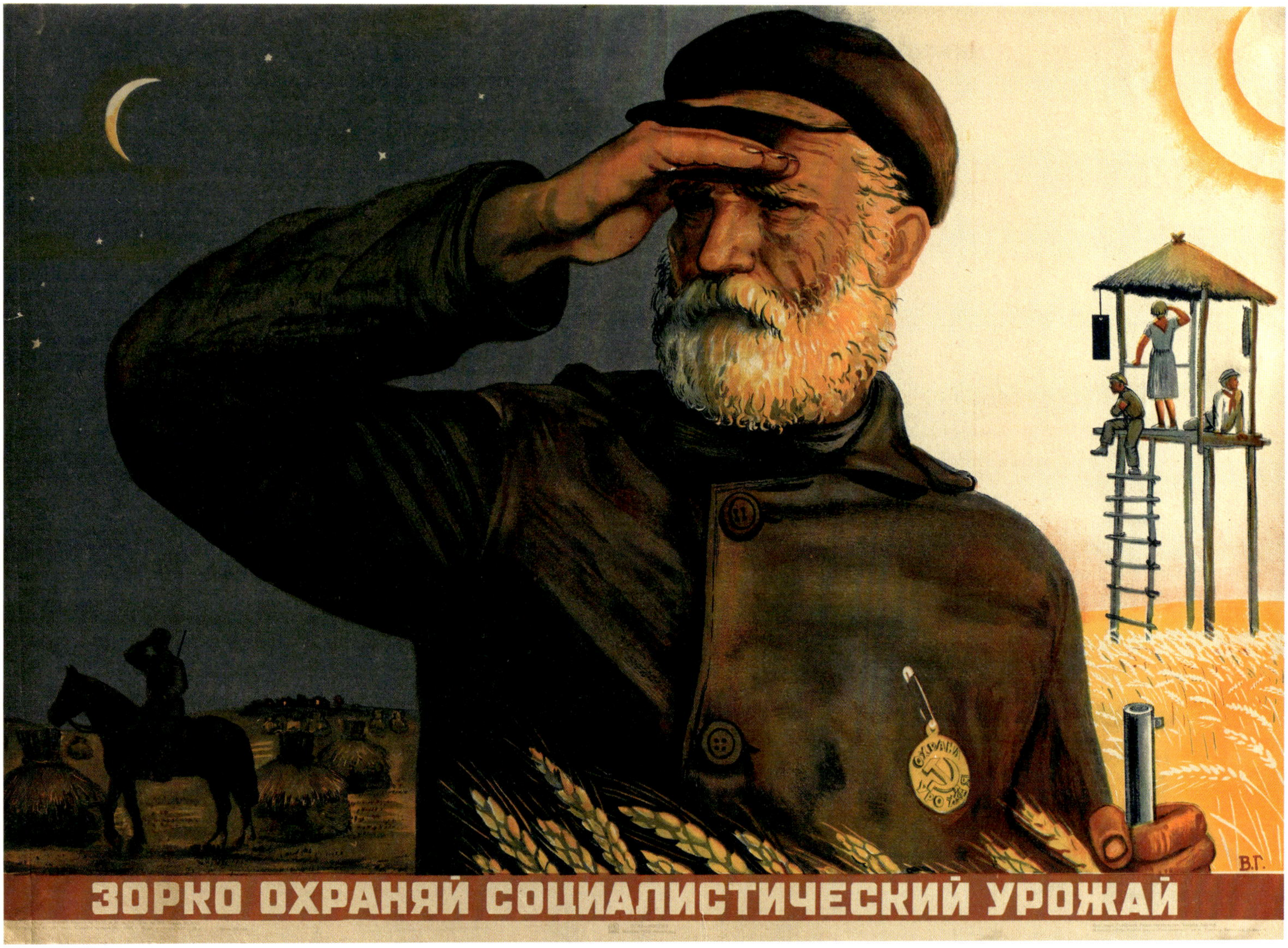

Fig. 36 V.I. Govorkov. "Vigilantly protect the socialist harvest." 1936.

with expulsion from the party for failing to step up and take the North Caucasus as a model of vigorous grain procurement. (On November 8, Stalin's wife, Nadezhda, committed suicide, following a marital spat; the loss devastated him.) Part of this campaign was surely an article in *Povolzhskaia Pravda* on November 11 by Pavel Postyshev (1887–1939), a senior party official and Stalin's close associate, who soon departed for the Volga region to press local officials to use "blacklisting" to squeeze out more grain. The article contended that going forward with "the class struggle and collectivization" was absolutely necessary, despite "temporary difficulties," for "the success of industrialization, which will make possible the strengthening of national defense and the reconstruction of agriculture."[23] Obviously, in the Bolshevik mind everything was interconnected—fighting against the "kulaks," implementing central planning, harsh repression, collecting grain, industrializing, building socialism.

In fact, in a speech to the Politburo and Central Committee Presidium on November 27, 1932, Stalin again blamed "kulaks" for sabotaging grain collection and demanded a "crushing blow" against them.[24] The Politburo enshrined the demand in a decree on December 14, including the threat of execution for "kulak sabotage," and armed brigades fanned out to rummage for grain.[25] Two days later, Goloshchekin argued that repression was unavoidable, since even collective farmers—alleged allies in class terms—were defying collection efforts.[26] In reality, opposition was pervasive. One secretary of the district party committee apparently asserted: "I will not walk over the bones of collective farmers in order to collect grain for the state."[27] Yet it was hard to resist pressure from above when the consequences could be dire. One collective farm chairman in Saratov Oblast, a party member since 1924, was arrested and imprisoned for four months because his farm was 4 percent shy of plan.[28] On December 23, local Soviets were authorized to determine the characteristics of "kulaks" depending on local conditions—a recipe for total arbitrariness.

The party authorities tightened control in town and city as well. A decree of November 15, 1932, stipulated that missing work even for one day, even with good reasons, could be grounds for termination and thus loss of ration tickets. Further, a decree of December 5 reorganized the ration system for workers to make access to rations fully dependent on employment. The law was draconian. The London *Times* reported that "attempts to utilize food-cards invalidated by the holder's dismissal for truancy will be punishable with death."[29] Often even the most dedicated workers, so-called *udarniki*, went hungry.

The leadership topped off its crackdown with a system of internal passports, for which Stalin pressed hard. Instituted on December 27, 1932, it was adopted for several reasons: to control migration, to keep track of labor, to filter out the wrong type of people ("kulaks, criminals, and other antisocial elements"),[30] to determine who should live in the major cities, and indirectly to tie down the peasantry, since peasants did not receive passports. They thus became second-class citizens.

At the end of December, the first Five-Year Plan was proclaimed "completed in four years." Also, with 73 percent of rural households collectivized, 1932 was declared the "year of the completion of universal collectivization."[31] Yet the results of both the plan and the campaigns were underwhelming, at best. True, farmers put 19 percent more land under the plow, but yields fell by about the same amount, and much grain was lost during harvest.[32] Only in 1958 did the country surpass the 1928 level of livestock output, and total draft power fell continuously throughout the

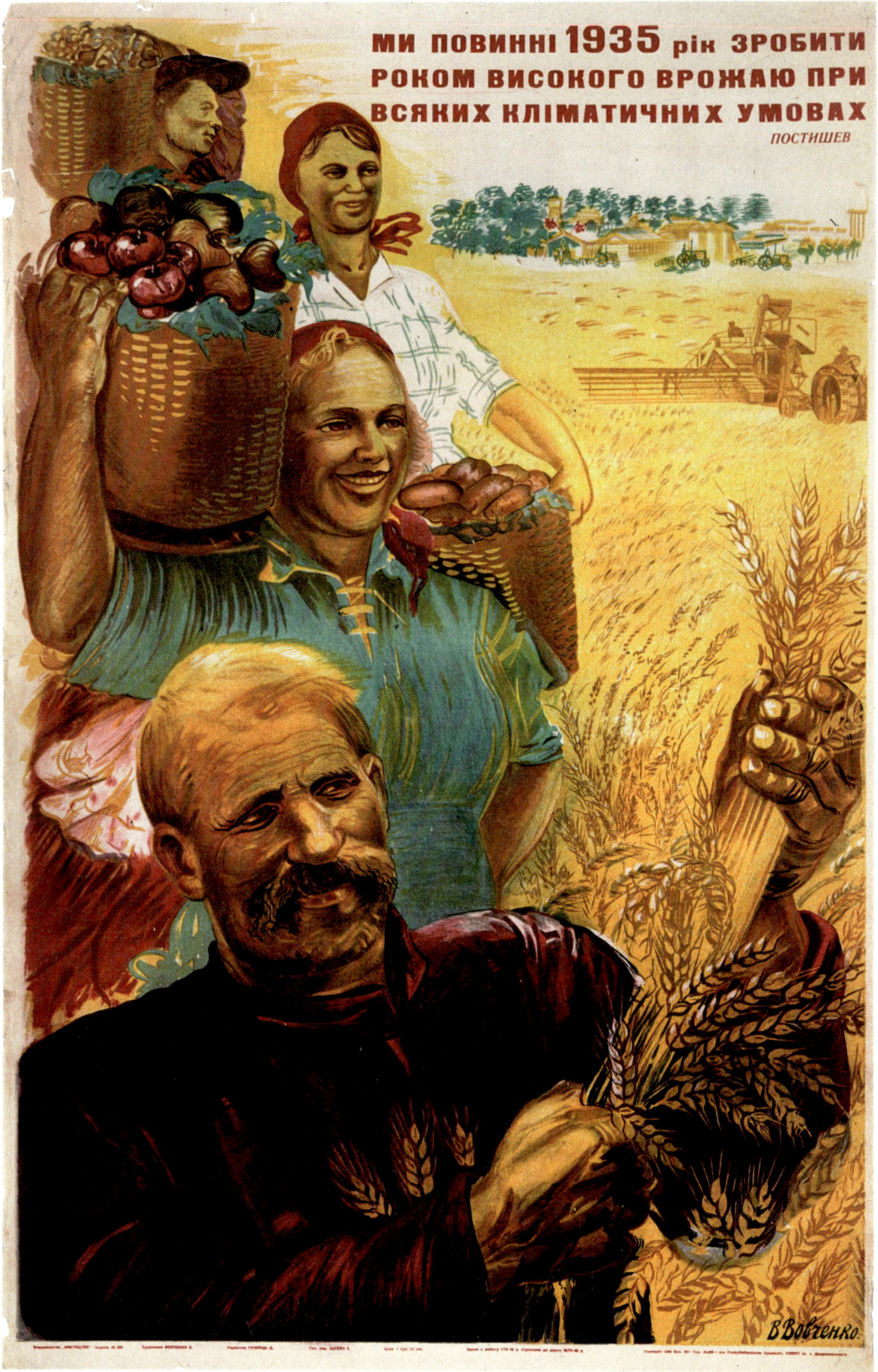

◀ **Fig. 37** V. Vovchenko. "We pledge in 1935 to produce a great harvest no matter the weather.—Postyshev." 1936. (In Ukrainian.)

▶ **Fig. 38** A.M. Kanarevsky. "Let us expand socialist competition and shock work in the collective farm fields." 1931. "We are entering into the 'Red List' Comrade ___ for the fulfillment and overfulfillment of the plan by ___%."

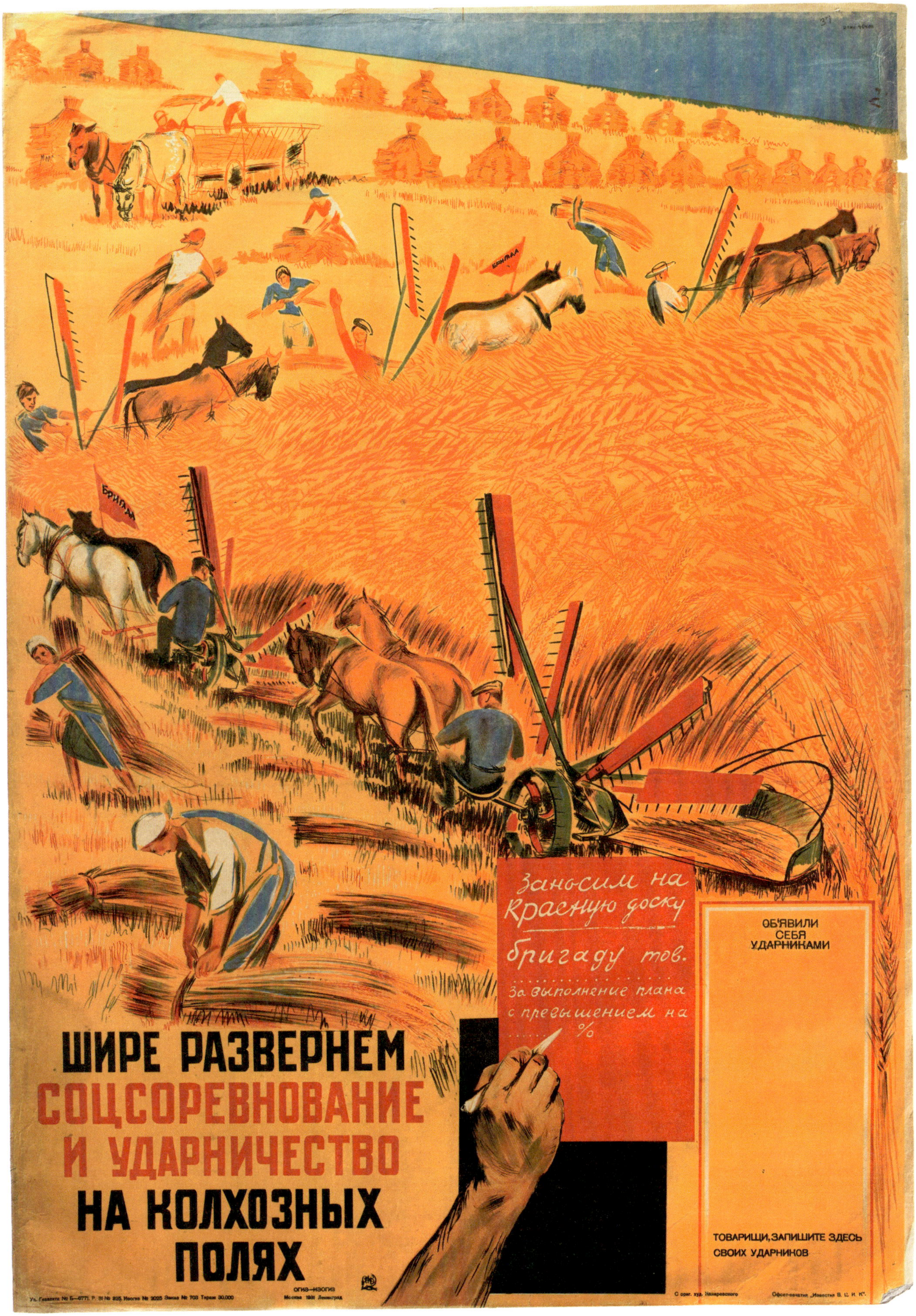

БРИГАДА
Заносим на
Красную доску
бригаду тов.
За выполнение плана
с превышением на
%
ОБ'ЯВИЛИ
СЕБЯ
УДАРНИКАМИ
ТОВАРИЩИ, ЗАПИШИТЕ ЗДЕСЬ
СВОИХ УДАРНИКОВ
ШИРЕ РАЗВЕРНЕМ
СОЦСОРЕВНОВАНИЕ
И УДАРНИЧЕСТВО
НА КОЛХОЗНЫХ
ПОЛЯХ
ОГИЗ—ИЗОГИЗ
Москва 1931 Ленинград

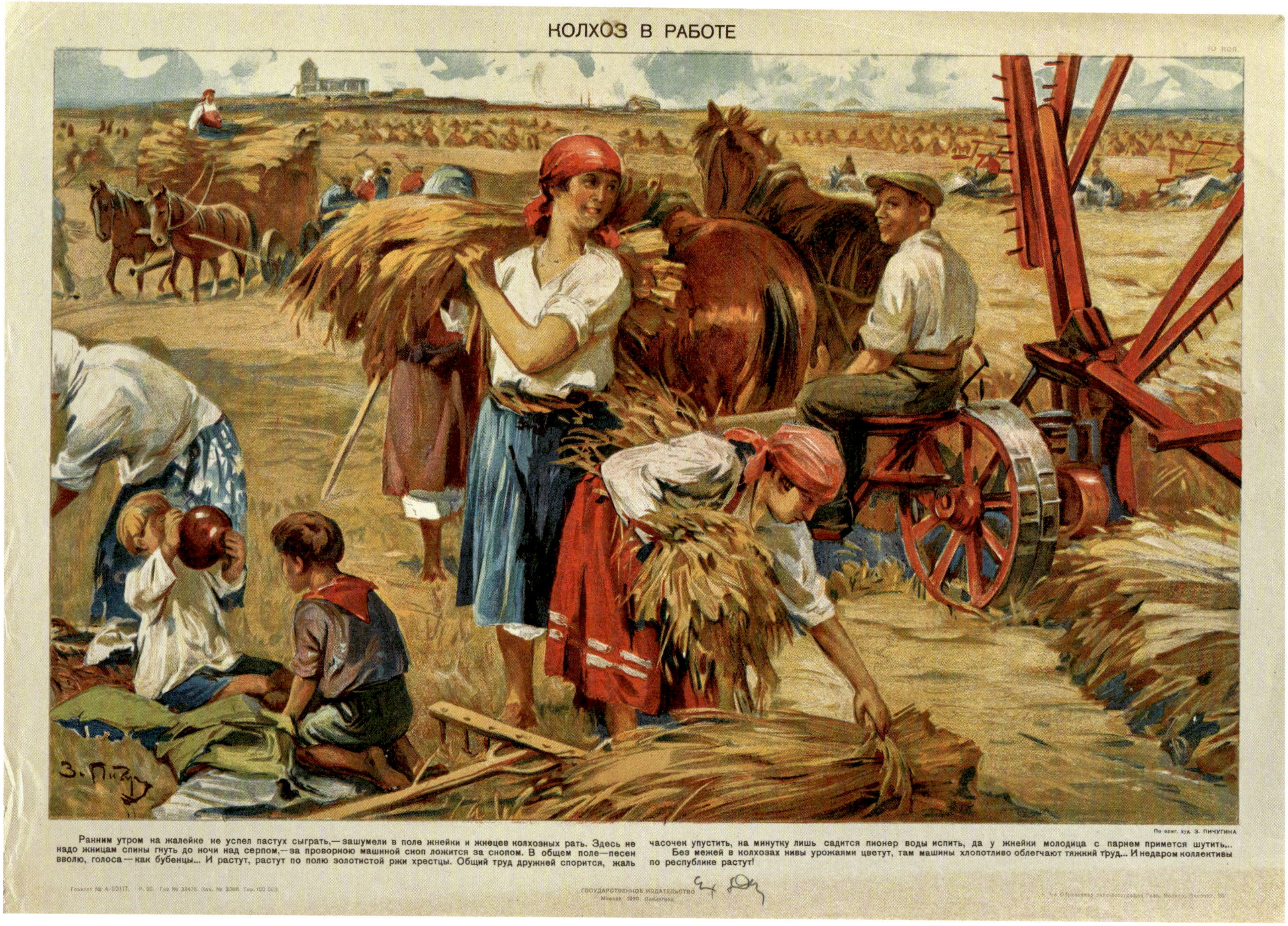

Five-Year Plan. Total agricultural output in 1932 reached only 107 percent of that of 1913.[33] Average daily consumption of grain, potatoes, lard, and butter in the countryside also declined over these years by 20–50 percent. The collections missed the plan target by 20 percent, and grain exports crashed from 5.18 million tons in 1931 to 1.8 million in 1932.[34] Tragically, a major famine struck the USSR in November 1932 and raged into June 1933. The biggest number of people perished in March–April. Only Moscow and Leningrad mostly escaped its ravages, thanks to large grain reserves.

The horrors people experienced during the famine are heart-wrenching. In all the major grain-producing regions, masses of ordinary people witnessed neighbors and family members consuming unhealthy ood substitutes and collapsing from hunger, children whining and begging for food, officials driving away unfortunates migrating in search of food, adults and children scavenging for something to eat, and kindred dreadful scenes of horror. One finds their harrowing stories in official reports, diaries, and letters to the editors of newspapers and to party leaders. An

Fig. 39 Z. E. Pichugin. "The collective farm at work." 1930. "In the early morning before the cock has even crowed, a throng of peasants is operating the threshing machine. . . ."

instructor at a technical institute in Ukraine brought his students to a collective farm to help with weeding and other chores. When they took a break for dinner, famished "children and even adults began to come running to beg for bread or *zatirka* [soup with nuggets of dough]. Struck by the horrible look of the hungry children, students began to give them pieces of bread and cooked *zatirka*. The children and adults hurled themselves greedily at the food. . . . Half an hour to an hour later, the children, who, after long starvation, had greedily eaten their fill of bread, began to fall to the ground, shouting and crying, writhing in pain." The children had not eaten regularly for so long that their stomachs could not tolerate even wholesome food. Later, the students offered some bread and hot porridge to a villager sent to sharpen their hoes. "The hungry man ate well," according to the instructor, "and about half an hour later he died right before our eyes."[35]

What is missing from the personal accounts and even official reports is the larger picture. For this, the best sources are by the few foreign journalists who dared to defy Soviet censorship restrictions. To do so meant expulsion from the country and prohibition to return. One such bold journalist was Gareth Jones (1905–35), a Welshman who served as a personal aide to former prime minister David Lloyd George and wrote for the Cardiff-based newspaper, *Western Mail*. Fluent in Russian, he traveled extensively in the USSR in 1930–33. His last trip, during the first two months of 1933, took him by foot and by rail through big swaths of Soviet countryside, where he conversed with hundreds of peasants. He also spoke to dozens of foreign diplomatic personnel, journalists, and technical experts. In a series of newspaper articles, he related the bitter suffering of the Soviet people. "Famine, far greater than the famine of 1921," he reported on April 8,

> is now visiting Russia. The hunger of twelve years ago was only prevalent in the Volga and in some other regions, but today the hunger has attacked the Ukraine, the North Caucasus, the Volga district, Central Asia, Siberia—indeed, every part of Russia. I have spoken to peasants or to eye-witnesses from every one of those districts and their story is the same. There is hardly any bread left, the peasants either exist on potatoes and cattle fodder, or, if they have none of these, die off.
>
> In the three agricultural districts which I visited, namely, the Moscow region, the Central Black Earth district, and North Ukraine, there was no bread left in any village out of the total twenty villages to which I went. In almost every village peasants had died of hunger.
>
> Even twenty miles away from Moscow there was no bread. . . .
>
> A little further on the road a woman started crying when telling me of the hunger, and said: "They're killing us. We have no bread. We have no potatoes left. In this village there used to be 300 cows and now there are only 30. The horses have died. We shall starve." Many people, especially in the Ukraine, have been existing for a week or more on salt and water, but most of them on beet, which was once given to cattle.[36]

Jones was denounced by Western journalists seeking to curry favor with the Stalinist regime, including the Pulitzer Prize–winning Walter Duranty (1884–1957), who wrote for the *New York Times*. Banned from visiting the Soviet Union, Jones turned his attention to the Far East. While traveling in Japanese-occupied Manchukuo in August 1935, he was murdered under mysterious circumstances. Some evidence later surfaced of Soviet secret police involvement.[37]

To what extent can the famine be blamed on poor weather? Viktor Kondrashin, who carefully studied meteorological data from

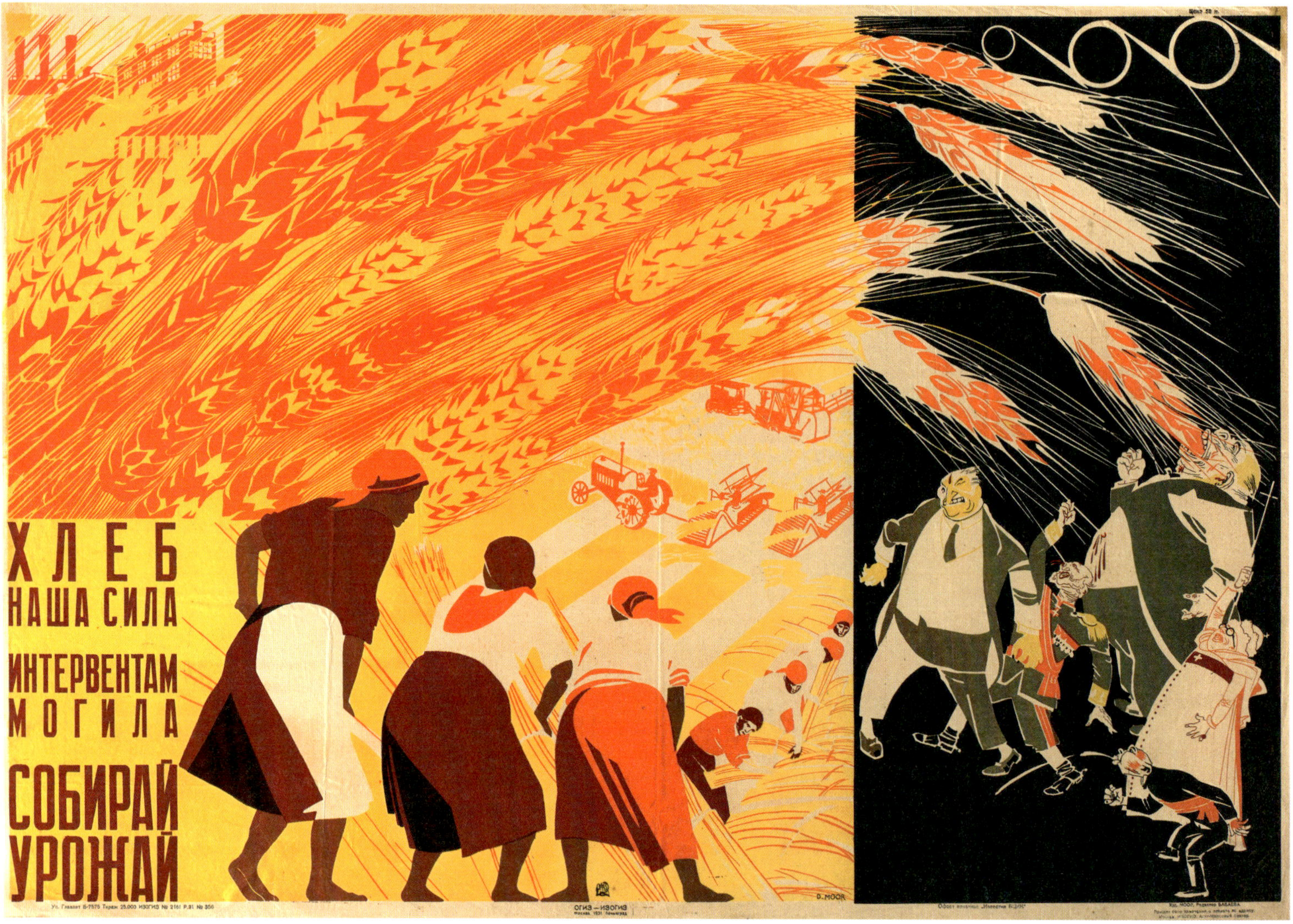

Fig. 40 D.S. Moor (Orlov). "Grain is our strength. To interventionists—the grave. Gather the harvest." 1931.

weather stations around the country and also interviewed hundreds of eyewitnesses, found evidence of some moderate dryness here and there in the Volga region but also determined that climatic conditions were mostly favorable in the chief grain-producing regions in 1932. "The famine was not caused by the will of God," he concluded, "but by the hand of man."[38]

Many government policies contributed to this outcome: the uprooting and deportation of millions of the country's most skilled farmers, heedless confiscation of foodstuffs without adequate compensation, the criminalization of the ancient practice of gleaning crop remnants in the fields, official discouragement of entrepreneurial agricultural work, the ideologically driven suppression of farmers' markets and the free exchange of produce, and the vilification of independent farmers. Still others were added in 1933. Most notorious and harmful was the prohibition instituted in January on the free movement of peasants seeking to flee famished regions. Stalin was convinced that the flight of peasants from hungry areas was nothing but counterrevolutionary activity. A central government directive of January 22 forbade the migration of peasants from the grain-producing countryside. Such activity, according to the directive, was being organized by enemies of Soviet power, by Socialist-Revolutionaries, and by agents of Poland with the goal of agitating "by means of the peasantry" against the collective farms and Soviet power.[39] By early March 1933, the political police had tracked down 219,500 peasants and had sent 186,600 back to their villages and sent the others to concentration camps.[40]

Stalin and the party leadership believed enemies were everywhere. At the January 7–12 party plenum, they voted to launch a party purge, which resulted in the expulsion from the party of eight hundred thousand members over the course of 1933.[41] They also decreed the establishment of a network of "political sections" (*politotdely*) at all the machine tractor stations (MTS) and state farms. Since November a commission had been working out the details. The plan was to appoint one thousand directors and two thousand assistant directors of political sections in Ukraine, the North Caucasus, and the Lower Volga. Their two main tasks were, first, to facilitate fulfillment by collective farmers of their obligations to the state and, second, to purge collective farms and MTS of class enemies. As Stalin warned the plenum attendees, "Today's kulaks and their helpers [*podkulachniki*] are for the most part quiet and even saintly. They hold positions throughout the collective farms and speak loudly in their favor. But they conduct such work of sabotage and undermining that the collective farms will not survive their work."[42] The political sections quickly got to work; within a few months they had rooted out seventy thousand collective farm officials, or 17.5 percent of their total staff.[43] Heavy pressure continued through the summer.

The regime also worked hard to conceal the grim reality or to deflect attention from it. On February 15–19, 1933, the first All-Union Congress of Collective Farm Shock Workers convened in Moscow. Kaganovich contrasted a hellish capitalist world where millions of workers and peasants were starving to the "glorious future" of the Soviet collective farm. (In early 1933, the press was full of bleak depictions of economic hardship in capitalist countries.) In his speech, Stalin emphasized the obligations of collective farmers. "Of you only one thing is demanded," he concluded, "and that is to work conscientiously; to distribute collective-farm incomes according to the amount of work done; to take care of collective farm property, to take care of the

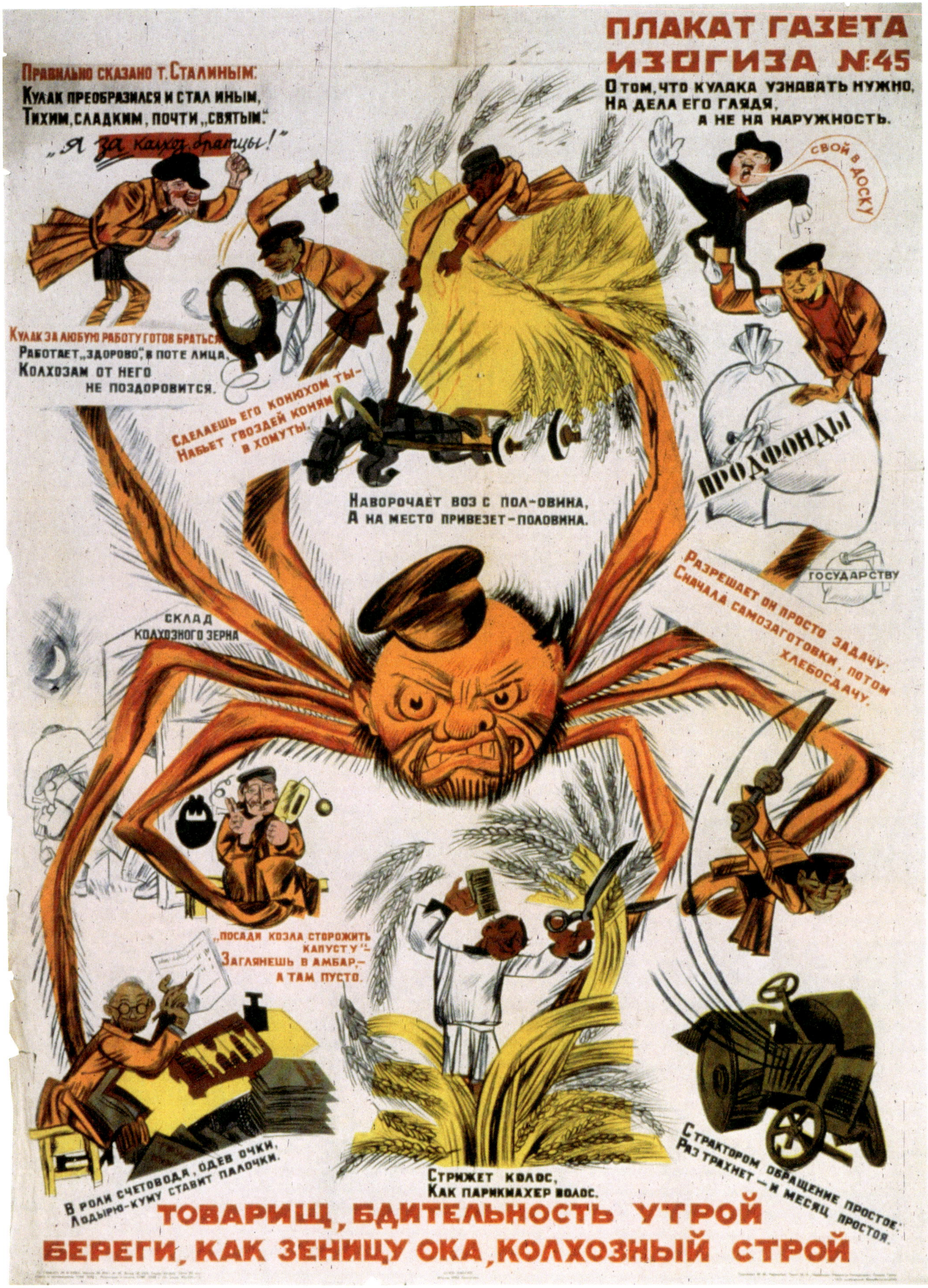
ПЛАКАТ ГАЗЕТА
ИЗОГИЗА №45
О том, что кулака узнавать нужно,
На дела его глядя,
а не на наружность.
Правильно сказано т. Сталиным:
Кулак преобразился и стал иным,
Тихим, сладким, почти „святым".
„Я за
Свой в доску!
Кулак за любую работу готов браться
Работает „здорово", в поте лица,
Колхозам от него
не поздоровится.
Сделаешь его конюхом ты —
Набьет гвоздей коням
в хомуты.
Наворочает воз с пол-овина,
А на место привезет — половина.
ПРОДФОНДЫ
ГОСУДАРСТВУ
Разрешает он просто задачу:
Сначала самозаготовки, потом
хлебосдачу.
СКЛАД
КОЛХОЗНОГО ЗЕРНА
„Посади козла сторожить
капусту" —
Заглянешь в амбар, —
а там пусто.
В роли счетовода, одев очки,
Лодырю-куму ставит палочки.
Стрижет колос,
Как парикмахер волос.
С трактором обращение простое:
Раз трахнет — и месяц простоя.
ТОВАРИЩ, БДИТЕЛЬНОСТЬ УТРОЙ
БЕРЕГИ, КАК ЗЕНИЦУ ОКА, КОЛХОЗНЫЙ СТРОЙ

tractors and machines; to see to it that the horses are well looked after; to fulfil the assignments of your workers' and peasants' state; to consolidate the collective farms and to expel from the collective farms the kulaks and kulak agents who have wormed their way into them."[44] In other words, they should devote themselves heart and soul to the state. Yet later in the speech, Stalin made an important concession: within a year or two, he promised, "there will not be a single collective farmer without his own cow. We Bolsheviks will see to it that every one of our collective farmers has a cow."[45] The fulfillment of this promise would take a long time, and its realization would depend more on the hard work of collective farmers than on state largesse, but limited individual ownership of livestock by collective farmers gradually became a cherished privilege and also an indispensable contribution to Soviet agricultural output. Nonetheless, the country was still nowhere near that "bright future." In fact, that very same day Stalin issued a ban on foreign journalists traveling freely in the country. Obviously there was still a lot to hide.

Repression continued as officials stripped the countryside of grain—roughly 40 percent of the entire harvest. Thousands of farmers were charged with sabotaging the sowing campaign. In January and the first half of February, nearly fifty thousand people had been sentenced, some four thousand to death, for theft of grain and other "socialist property" in fulfillment of the August 7, 1932, law.[46] The central authorities scapegoated collective farm officials for the disastrous conditions, beginning in late February. On March 11, the OGPU publicly announced it had summarily executed thirty-five directors, bookkeepers, and other collective and state farm officials for their responsibility in the "'catastrophic' losses of tractors and other agricultural machinery." They were accused, as reported by the London *Times*, of belonging "to a wrecking organization of officials of non-proletarian origin bent on reducing conditions in the U.S.S.R. to starvation."[47]

On March 27, the Soviet press denounced German, American, Italian, and other manufacturing firms for delivering supposedly defective industrial equipment with the intention of "wrecking" Soviet industrial progress. *Pravda* and other newspapers, as reported by the London *Times*, warned that the leadership was "obliged to fight on all fronts inside and outside the party. Some strata of the industrial masses lately show signs of wavering and the entire peasantry is polluted with kulak ideas." To bring these elements into line, the government was sending "detachment after detachment, composed of thousands of the staunchest proletarians and headed by the most experienced revolutionary leaders."[48]

Local officials often used brutal methods. In a letter to Stalin on April 4, Sholokhov revealed abuses that had come to his attention in the Veshensky district of the North Caucasus region, where he lived. Some 1,090 families (from a total of 13,813 in the district) were turned out of their homes in the freezing cold, many for having given shelter to exiled "kulaks." In one village, Sholokhov personally saw "at night, under a fierce wind, in freezing weather, when even the dogs hide from the cold, families, who were turned out of their homes, lighting fires in the alleys and sitting next to them. Children wrapped themselves in rags and lay on the semi-frozen ground, their screams filling the air." He knew of one instance when a woman with an infant was turned out of her collective farm. "All night she went from house to house, begging to be let in with her infant. All refused out of fear that they too would be turned out. By morning the baby had frozen stiff in the arms of its mother." In all, Sholokhov

Fig. 41 M.M. Cheremnykh. "Comrade, vigilance at all times. Guard the collective farm like the apple of your eye." Plakat-gazeta IZOGIZa N° 45. 1933. "Today's kulak has become 'quiet and even saintly.'—Stalin."

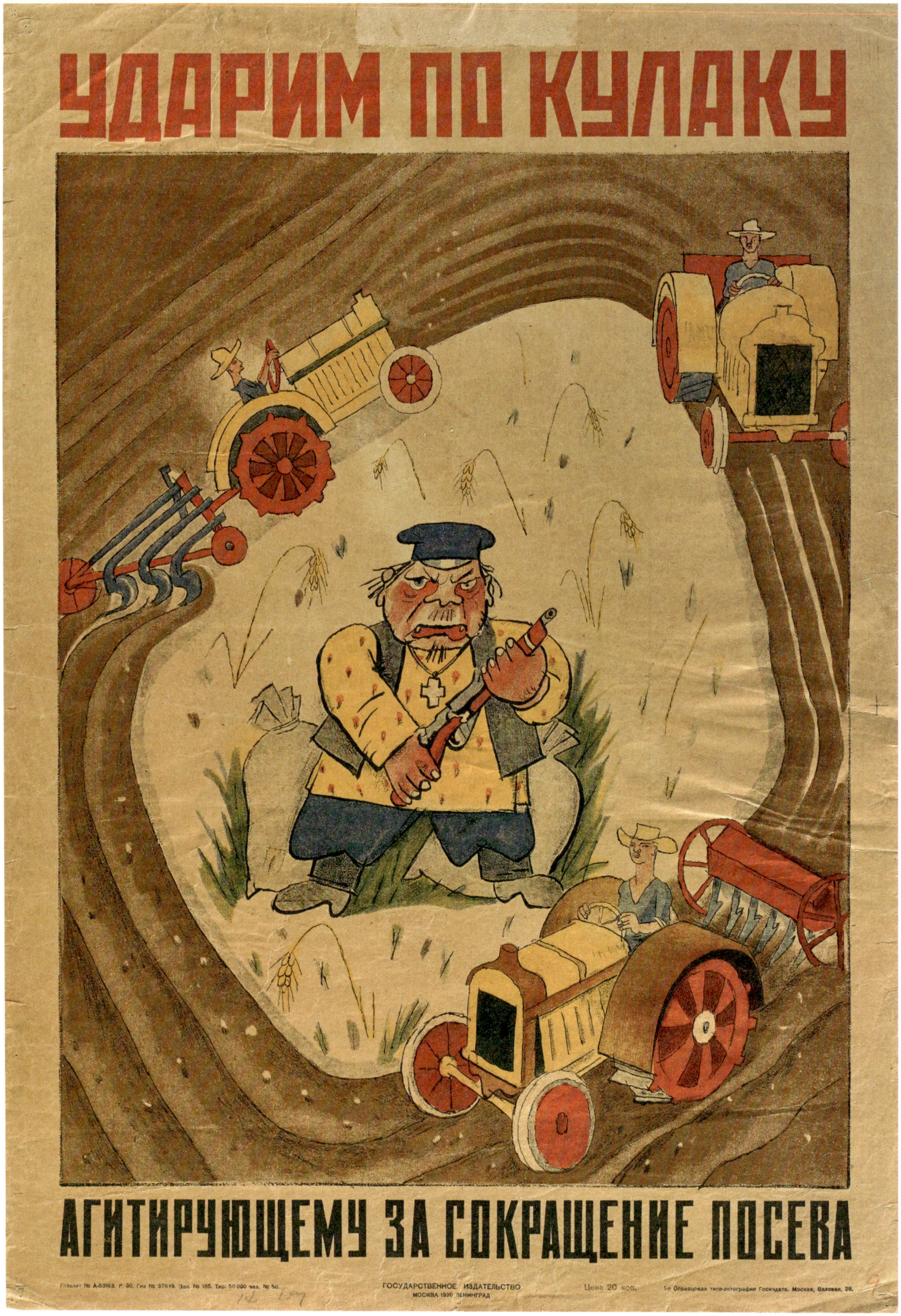
УДАРИМ ПО КУЛАКУ
АГИТИРУЮЩЕМУ ЗА СОКРАЩЕНИЕ ПОСЕВА
Главлит № А-53163. Р. 30. Гиз № 37849. Зак. № 185. Тир. 50 000 экз. № 50.
ГОСУДАРСТВЕННОЕ ИЗДАТЕЛЬСТВО
МОСКВА 1930 ЛЕНИНГРАД
Цена 20 коп.
1-я Образцовая типо-литография Госиздата. Москва, Валовая, 28.

listed seventeen methods of torture, such as locking half-naked people in freezing outbuildings, firing guns close to peoples' faces, making them stand barefoot on a red-hot stove, dousing women's legs and skirts with kerosene and briefly lighting it, and conducting interrogations in the presence of a corpse. "Such examples," he added, "could be multiplied without end."[49]

Stalin wrote back promising food aid for the hungry in that district and admitting that the abuses were not justified, but added that Sholokhov was only seeing one side of the problem: "The respected grain growers of your district (and not only yours) were engaged in work slowdowns (sabotage!) and were not averse to leaving workers, the Red Army without bread. The fact that the sabotage was quiet and outwardly inoffensive (bloodless), this fact does not change the fact that the respected grain growers have in reality been waging a 'quiet' war with the Soviet regime. War by attrition, dear Comrade Sholokhov."[50] Food aid was in fact delivered, and some of the culprits named by Sholokhov were punished, though none particularly severely. How many brutal local officials were not denounced by a whistle-blower of the stature of Sholokhov? Moreover, if Stalin really cared about insufficient food for workers and soldiers, why did the regime export 354,000 tons of grain in the first half of 1933 (and over 1.3 million in the second half),[51] which was just over the total amount of food aid (320,000 tons) distributed during the same period?[52] As Nicholas Werth has pointed out, however, with 30 million starving people, that worked out to just over 10 kilograms per person.[53] Assuming it had to last them for at least four months, until summer vegetables began to ripen, the aid amounted to only a little over 3 ounces per person per day.

Fig. 42 Artist not identified. "Strike the kulak, who agitates for reduced sowing." 1930.

In March and April, senior officials received a steady stream of reports of starvation. Millions of peasants were debilitated from hunger. Many adults age twenty to forty were dying because they had been giving their food to their children. Cannibalism was rife. In the epicenters of famine, children were typically the first victims; their crazed parents or near relatives, the perpetrators. The streets of such areas became full of children age three to fourteen years. Many parents abandoned them at official offices or train stations. One director of an MTS political section opined in this regard that "kulaks are deploying a new subterfuge: they abandon their children and then abscond."[54] Nor did all officials take the famine seriously. The party boss of Vinnitsa Oblast in west-central Ukraine dismissed such reports in April 1933, arguing in part that "those who are starving are loafers, parasites who expend great energy disorganizing the collective farms."[55] Of course, the central authorities knew full well that a demographic catastrophe was under way; why else forbid the recorders of vital statistics (ZAGS) to mention famine in official death records?[56] (In 1934, moreover, the death registers for 1932–33 were apparently systematically destroyed.)[57]

Amid the worst famine in recorded Russian history, the country turned a corner: the regime triumphed. On April 27, 1933, a party official reported to Iosif Vareikis (1894–1938), the party boss of the Central Black Earth Region, on terrible local conditions, including high mortality, thousands distended from hunger, and the mass abandonment of children. "Despite these grave conditions in the district," he concluded, "collective farmers are taking measures to plant their fields so as in the future to avoid" such conditions.[58] Their resistance, in other words, had been broken. Stalin and Molotov soon reached a similar conclusion. In a secret directive of May 8, 1933, they wrote to all party and Soviet officials that the years of collectivization had been characterized by an intensification of

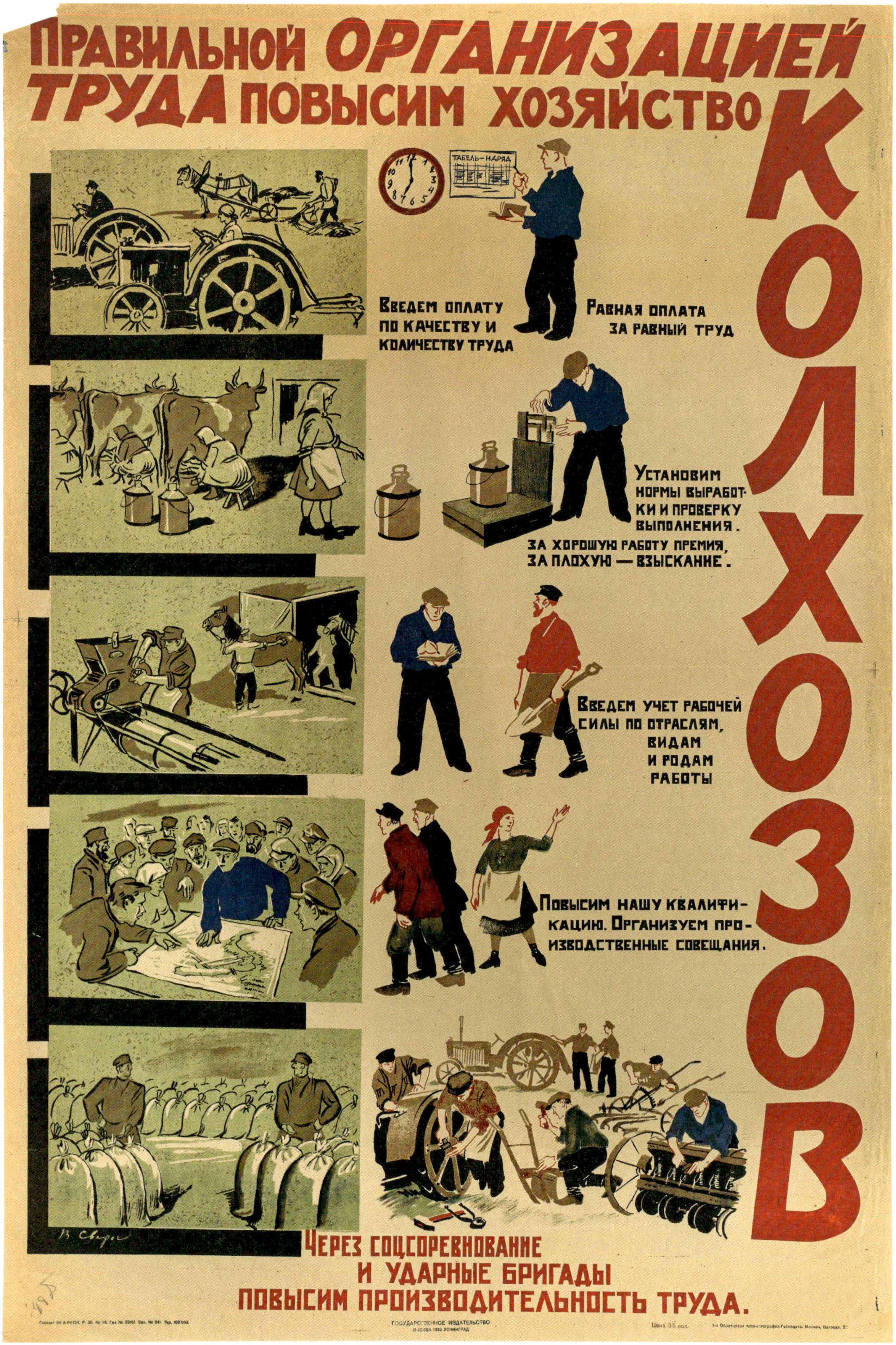
Правильной ОРГАНИЗАЦИЕЙ
ТРУДА повысим хозяйство
КОЛХОЗОВ
Табель—наряд
Введем оплату по качеству и количеству труда
Равная оплата за равный труд
Установим нормы выработки и проверку выполнения.
За хорошую работу премия, за плохую — взыскание.
Введем учет рабочей силы по отраслям, видам и родам работы
Повысим нашу квалификацию. Организуем производственные совещания.
Через соцсоревнование и ударные бригады повысим производительность труда.
В. Свирин
Государственное издательство
Москва 1930 Ленинград
Цена 35 коп.

◀ **Fig. 43** V.S. Svarog (Korochkin). "Through the correct organization of labor, we will raise the productivity of the collective farms." 1930. "Wages will be determined by the quality and quantity of labor. . . . For good work, prizes; for inferior, punishment. . . ."

the class struggle in the countryside "against kulaks, thieves, and all sorts of saboteurs." They continued, "Three years of struggle have resulted in the destruction of the forces of our class enemies in the countryside." Thus, "a more favorable conjuncture" had been reached, permitting "for the most part the curtailment of the application of mass exile and harsh forms of repression."[59] The leadership felt in control to such an extent that on June 8, 1933, it ordered the courts to halt the mass application of the August 7 law. (In all, the application of the law had resulted in 9,163 executions and 152,908 people sentenced to ten years' "deprivation of freedom."[60] An analysis of twenty thousand such cases qualified only 15 percent as "well-to-do.")[61] The Soviet writer Alexander Serafimovich (1863–1949) surely spoke for the regime when he declared on August 9 that the entire peasantry admitted its guilt for stealing grain before the proletarian government.[62] Or as Khataevich told Viktor Kravchenko, an official charged with grain collection in early 1933 who later defected to the United States,

▼ **Fig. 44** Artist not identified. "Muslim peasants joining the collective farm." 1939–40?

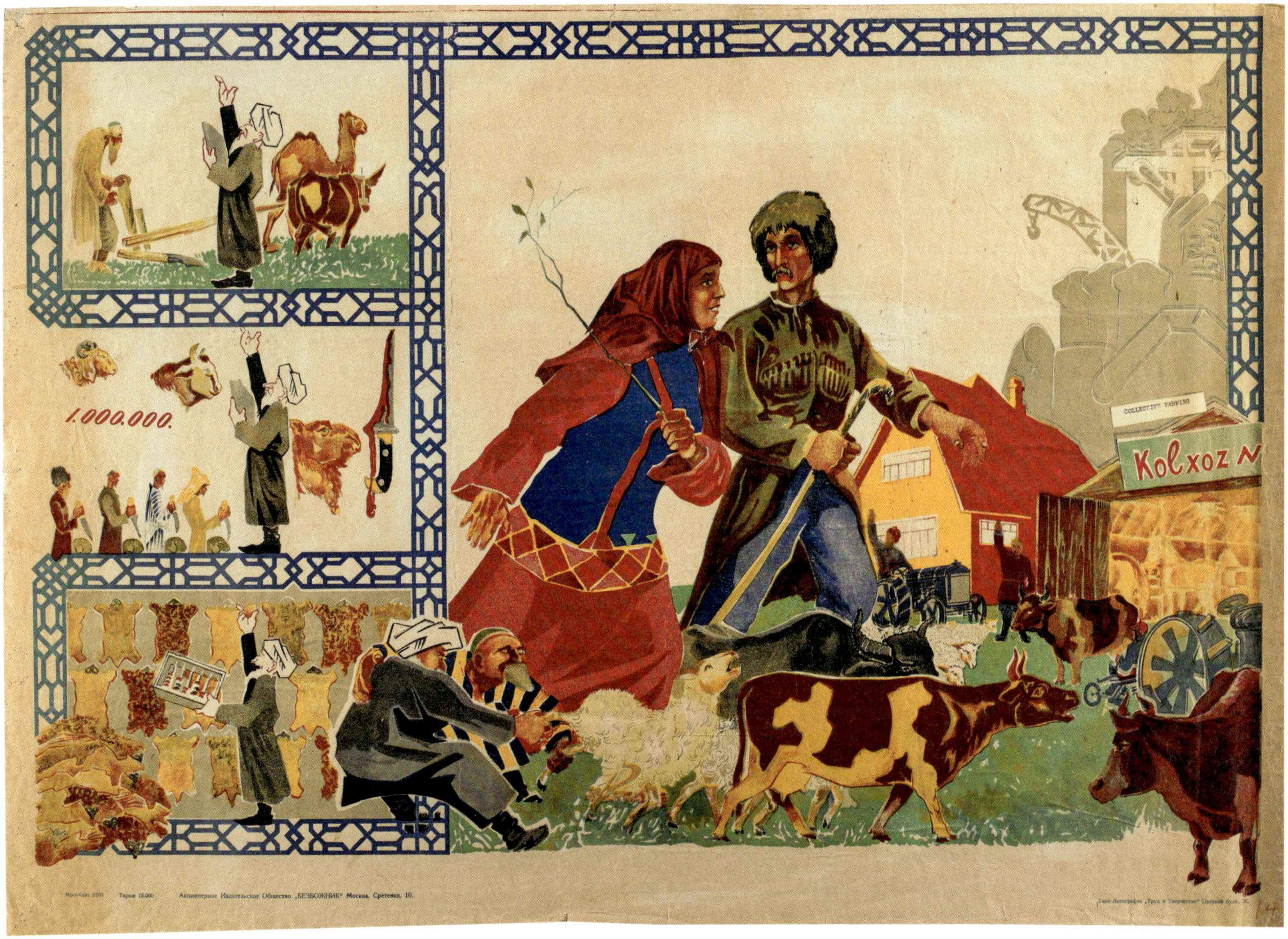

"A ruthless struggle is going on between the peasantry and our regime. It's a struggle to the death. This year was a test of our strength and their endurance. It took a famine to show them who is master here. It has cost millions of lives, but the collective farm system is here to stay. We've won the war."[63] Privately, he admitted feeling sympathy for the suffering peasantry, but overtly he gave full lip service to the foul official line.

The destruction wrought by the famine was unparalleled. The number of "excess deaths" throughout the country is still disputed, but nearly all scholars would reckon the population losses (including babies never born because of the harsh conditions) to have been at least 5.5 million and more likely closer to 7 million—or even higher if one includes the many hundreds of thousands of people who permanently fled the country, mostly from Kazakhstan, because of the famine.[64] Most of the deaths occurred in rural areas. Some 429,000 were "kulak" special settlers.[65] The regions that suffered the most were Ukraine, which lost between 3.5 and 4.5 million, and Kazakhstan, where some 1.5 million died from starvation, or approximately 35 percent of the population. According to Viktor Kondrashin's calculations, based upon work in the archives of sixty-five registry offices, along with many other sources, some two hundred thousand to three hundred thousand people died from famine in the Volga and southern Ural regions in 1932–33.[66] Hundreds of thousands more died in the North Caucasus and the Central Black Earth Region.[67]

Despite such catastrophic losses, the famine officially never occurred. It was not mentioned in the press. Soviet officials breathed not a word. Tens of thousands of party members, who had enjoyed access to copious rations and the network of special closed stores, may themselves have had no idea of the suffering of so many millions throughout the country.

Scholars have argued about whether the famine, particularly in Ukraine, constituted the intentional destruction of a people, that is, genocide. It seems quite certain that Stalin and the rest of the leadership, as well as many officials up and down the chain of command, intended to teach a lesson to the peasantry throughout the country, to show them who was boss. It is also obvious that more Ukrainians than other Soviet peoples suffered from the famine and the policies that chiefly caused the famine, including disastrously large grain collections, prohibiting peasants from fleeing the hungry regions, and a further 268,091 "kulaks" exiled in 1933. But statistically, a dramatically larger proportion of Kazakhs—roughly one in five—perished from famine in 1932–33 than did Ukrainians (roughly one in ten). Nor did the authorities permit any foreign deliveries of food aid to the starving, which Western governments and NGOs were eager to supply. Clearly Stalin and his supporters in the government did not wish to make public the catastrophe to which they had driven the country. Yet, in 1933 the government gradually curtailed procurements and sent modest food aid to Ukraine and Kazakhstan—mostly, to be sure, in the second half of the year, after the greatest suffering had occurred. A lot depended, moreover, on the party bosses in the regions. It seems that Postyshev, who was sent to the Volga region, did not push as hard for grain collection as Molotov and Kaganovich, who were sent to Ukraine and the North Caucasus.

For the rest of the year, grain on collective farms was provided only to those who went into the fields and worked, and it was provided only day by day. Officials saw it as a weapon for defeating "kulaks"

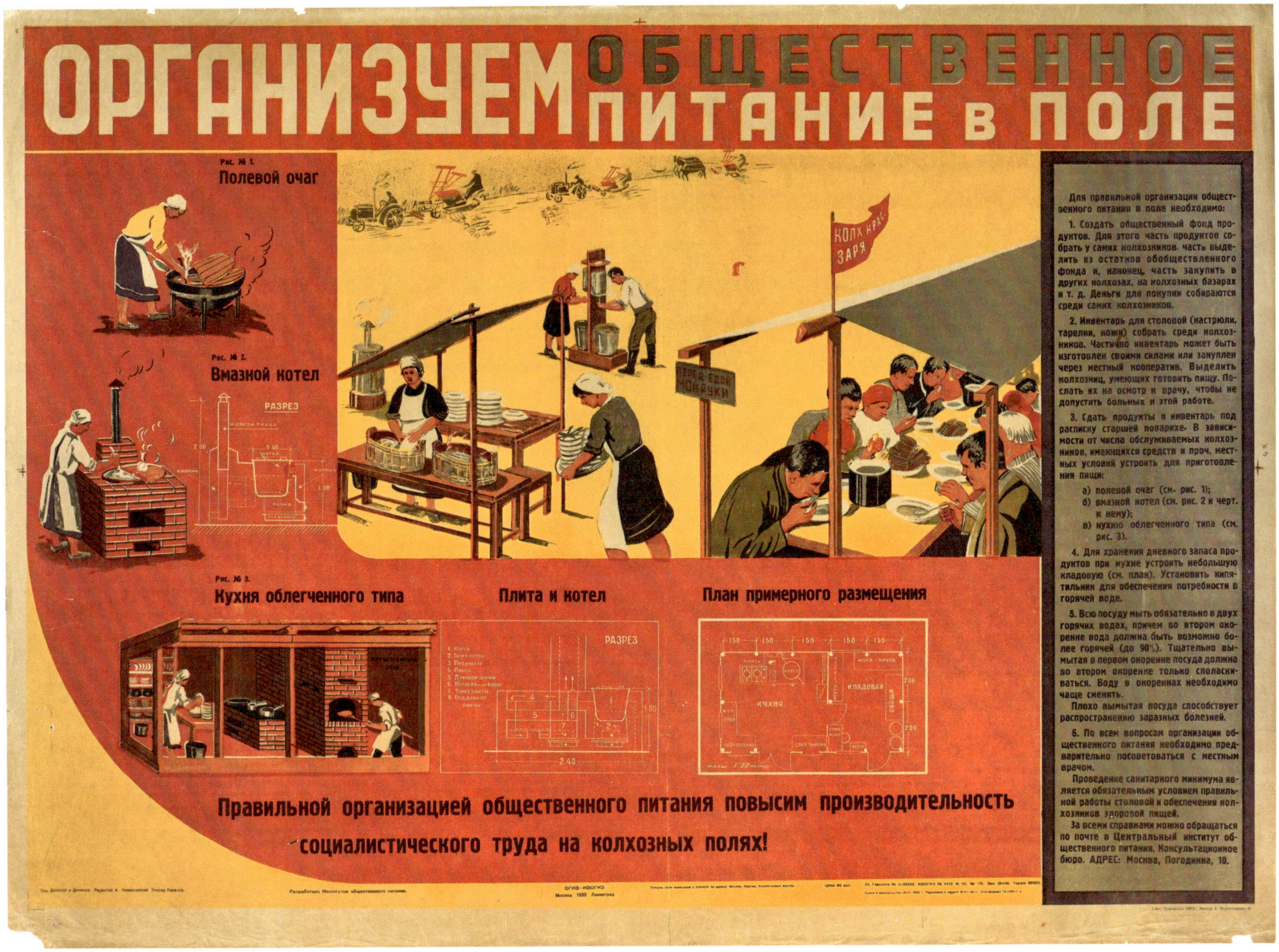

Fig. 45 I.I. Donskoi. "We will organize socialized nourishment in the fields." 1932. "Through a correct organization of socialized nourishment, we will raise the productivity of socialist labor in the collective farm fields!"

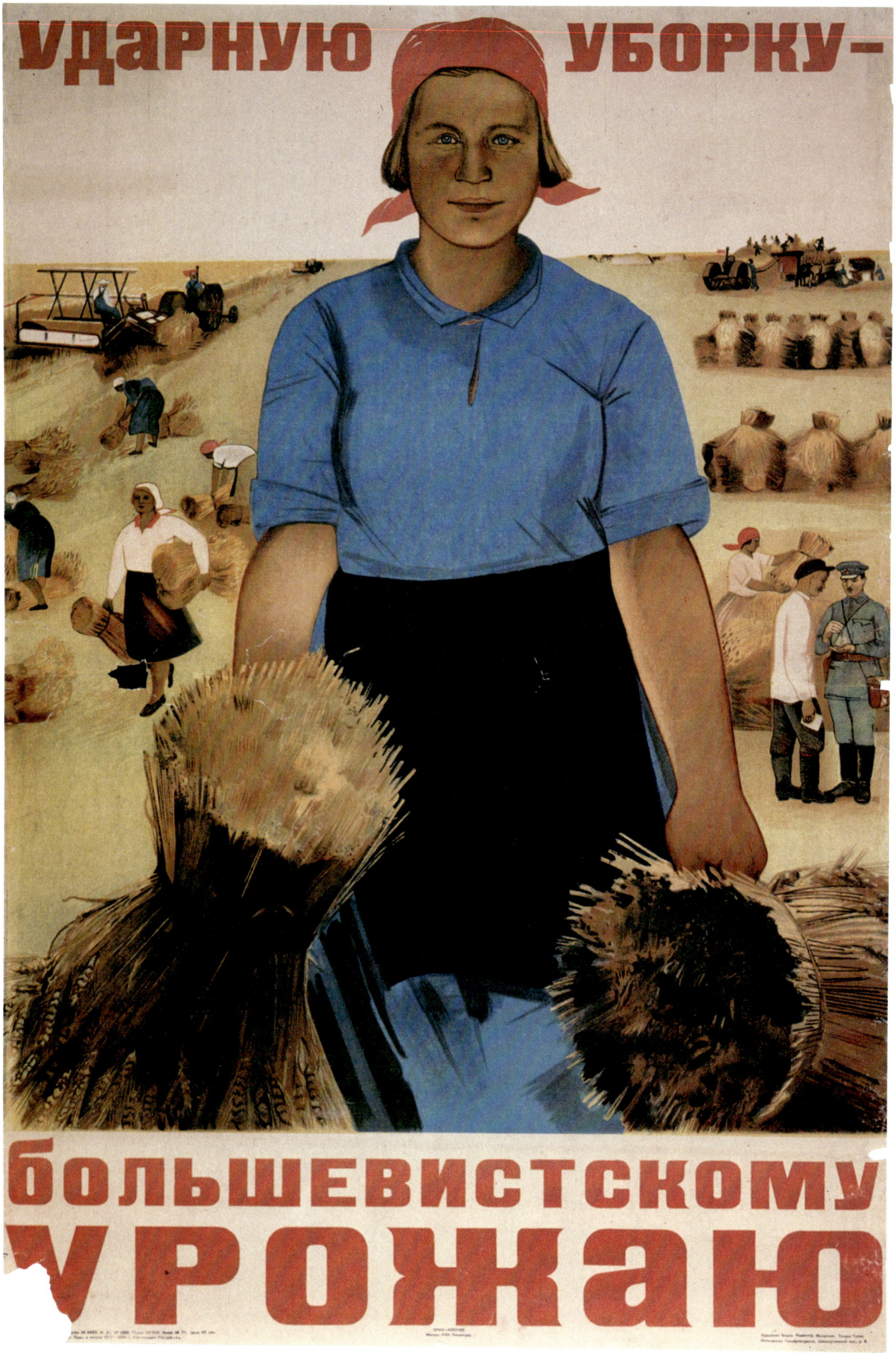
Ударную уборку —
большевистскому
урожаю

Fig. 46 M.A. Voron. "An energetic gathering of the Bolshevik harvest." 1934.

and an instrument for fulfilling the plan.[68] Those who did not work (except non-able-bodied family members of Communists) received little or none at all. The secret police reported from around the country that summer that peasants who had previously dragged their feet now worked eagerly. Largely as a result of the change in peasant attitude, by October 1 the grain collections had yielded 16.4 million tons—almost twice as much as in 1932.[69] A month and a half later, the goals of the grain collection plan had been achieved—earlier than any year since the beginning of central planning.

Yet even this quantity fell short of the country's needs, although the government had halted its export program in July, and in October had contracted to import 500 tons of high-grade winter wheat from Canada.[70] But most of this grain was earmarked for the cities. Shortages and even instances of famine in the countryside, including on collective farms, continued until summer 1934. Hardship in town and country was caused also by the unprecedented decline of livestock. The number of cattle, horses, and pigs dropped by roughly 50 percent from 1928 to 1933, sheep and goats by nearly 300 percent.[71] The number of cattle did not climb back to the 1914 level until around the time the Soviets sent the first man-made satellite (*Sputnik*) into orbit.[72]

Collective farm households began to produce comparatively large amounts of agricultural products on their individual plots of land, access to which was enshrined in a directive of December 23, 1933. During that year, they produced 61.2 percent of the potatoes and 72 percent of the vegetables produced on the collective farms.[73] Most of the meat they consumed was also of their own production. In December, the government purchased sixteen thousand workhorses to be distributed to collective farmers in Ukraine, Belorussia, and Western Region.[74] By the end of the next year, nearly two-thirds of all collective farm families possessed a cow. Gradually, the collective farmers, when tending to their gardens and their personal livestock, became the most productive agricultural workers in the country. From now on, the official policy would exclude the use of violent coercion and pressure to collect grain characteristic of the collectivization drive through the first half of 1933, though here and there some officials were not listening and continued to apply the old methods. Even so, a somewhat uneasy equilibrium was achieved. The regime was happy because it received bigger procurements than ever before, and the rural dwellers were relieved that most officials were no longer bleeding them dry.

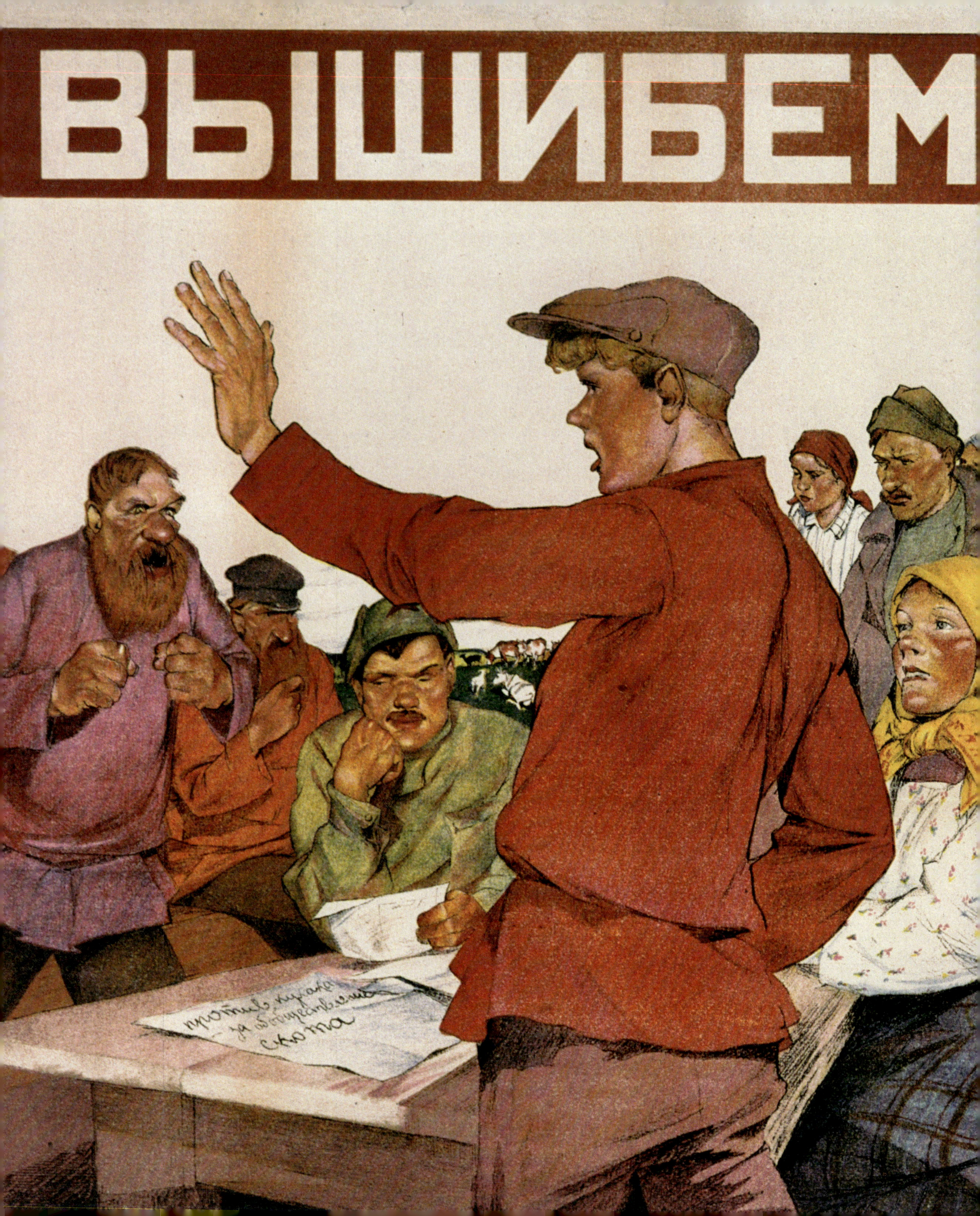
ВЫШИБЕМ
против кулака
— за обобществление
скота

Chapter 4

A BROKEN PEOPLE, 1934–

The terrible hardship imposed by the famine broke the resistance of the Russian peasantry, which for centuries had risen up against the central authorities. Henceforth, it would remain docile. This chapter will draw upon a wealth of propaganda posters portraying many of the goals and policies that the Soviet leadership pursued in the countryside. This is the story of revolutionary intentions and dreams. Among the themes embodied in numerous posters available in the Hoover Archives are socialist competition, the regimentation of society, centralized planning, the mechanization of farming, innovative agricultural techniques such as seed selection, the rejection of traditional methods of husbandry, rapid production tempos, social mobility from farm to factory, endless mobilization campaigns, gender equality, increasing literacy, the solidarity of rural and urban citizens, and the importance of contributing resources to the industrialization of the country. These policies taken together were intended to dramatically raise agricultural output, to contribute to fashioning the "new Soviet person," and to help transform the country into a modern, industrialized economy.

The Seventeenth Party Congress, held in January 1934, the "Congress of Victors," proclaimed that socialism had been built in the USSR. In his report to the congress on the work of the Central Committee, Stalin asserted that

> during this period, the U.S.S.R. has become radically transformed and has cast off the aspect of backwardness and medievalism. From an agrarian country it has become an industrial country. From a country of small individual agriculture it has become a country of collective, large-scale mechanised agriculture. From an ignorant, illiterate, and uncultured country it has become—or rather it is becoming—a literate and cultured country covered by a vast network of higher, secondary and elementary schools functioning in the languages of the nationalities of the U.S.S.R.

He added later in the speech that 65 percent of all peasant households in the country had been collectivized. All that remained was "the gradual absorption and re-education of the remaining individual peasant farms and farmers by the collective farms. This means that the collective farms have triumphed completely and irrevocably." The audience erupted, as the cliché had it, in "stormy and prolonged applause."[1]

Yet in the collective farms and the countryside generally all was not well. On April 26, 1934, the newspaper *Pravda* complained of many instances where in

◀ Detail of "We will kick the kulaks out of the collective farms," *see page 106.*

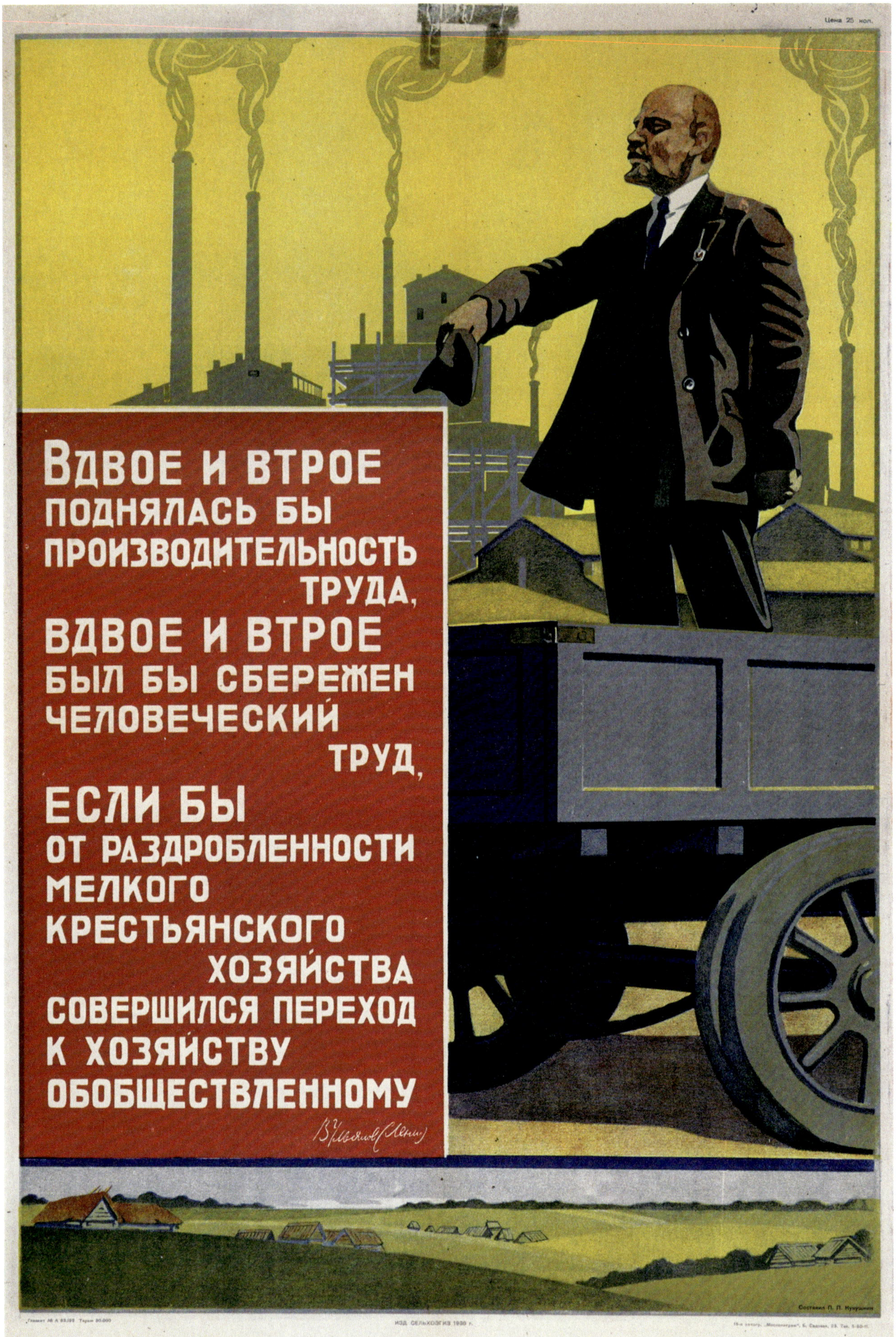
Цена 25 коп.
ВДВОЕ И ВТРОЕ
ПОДНЯЛАСЬ БЫ
ПРОИЗВОДИТЕЛЬНОСТЬ
ТРУДА,
ВДВОЕ И ВТРОЕ
БЫЛ БЫ СБЕРЕЖЕН
ЧЕЛОВЕЧЕСКИЙ
ТРУД,
ЕСЛИ БЫ
ОТ РАЗДРОБЛЕННОСТИ
МЕЛКОГО
КРЕСТЬЯНСКОГО
ХОЗЯЙСТВА
СОВЕРШИЛСЯ ПЕРЕХОД
К ХОЗЯЙСТВУ
ОБОБЩЕСТВЛЕННОМУ
Составил П. П. Кукушкин
ИЗД. СЕЛЬХОЗГИЗ 1930 г.

◀ **Fig. 47** P.P. Kukushkin. "You will double and triple your productivity, you will double and triple your labor efficiency, if you abandon your scattered small landholdings and enter a collective farm. —Ulianov [Lenin]." 1930.

▶ **Fig. 48** D.S. Moor (Orlov). "We were a country of the wooden plow. We have become a country of the tractor and the combine. —Kaganovich." 1934.

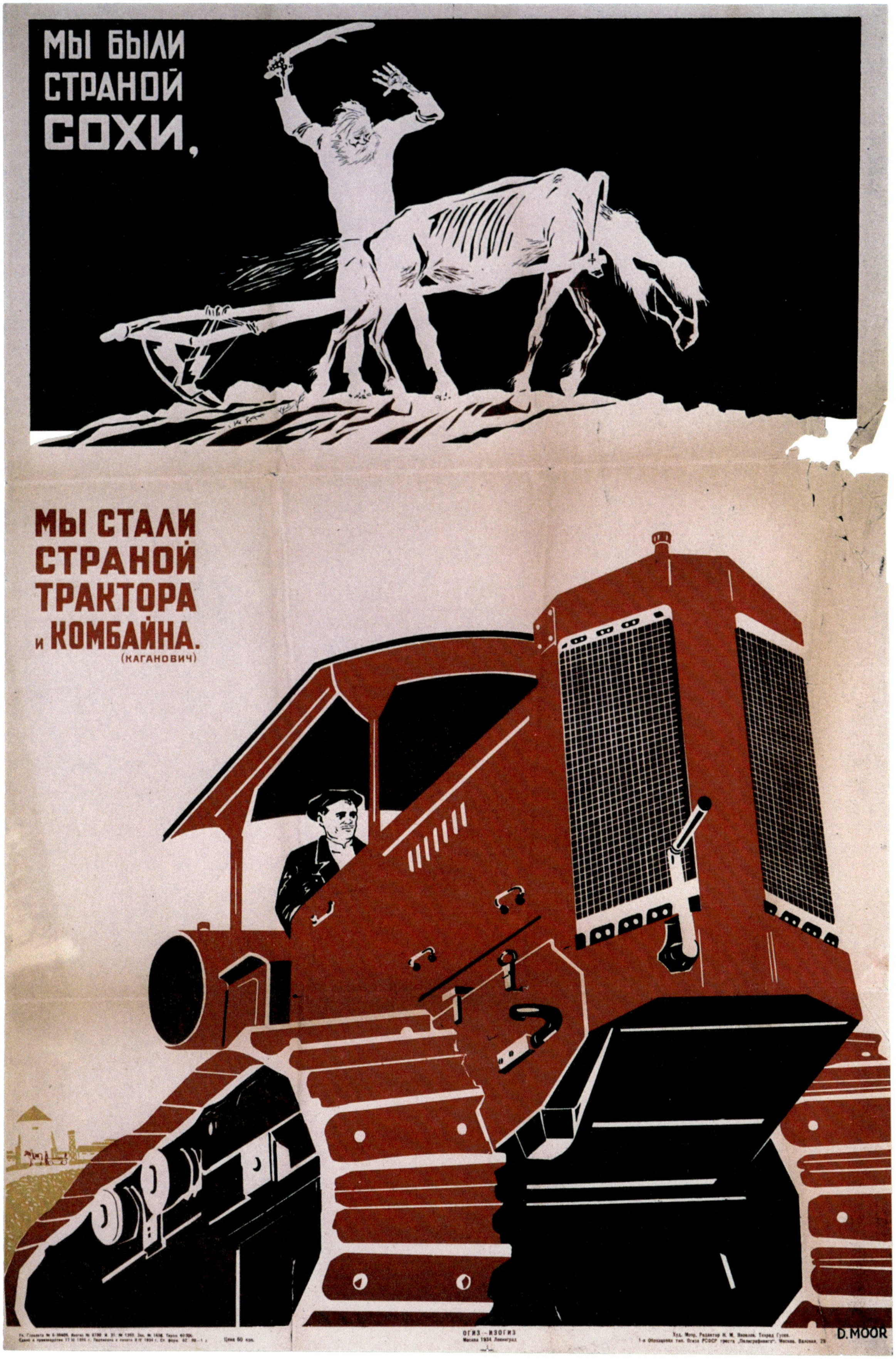

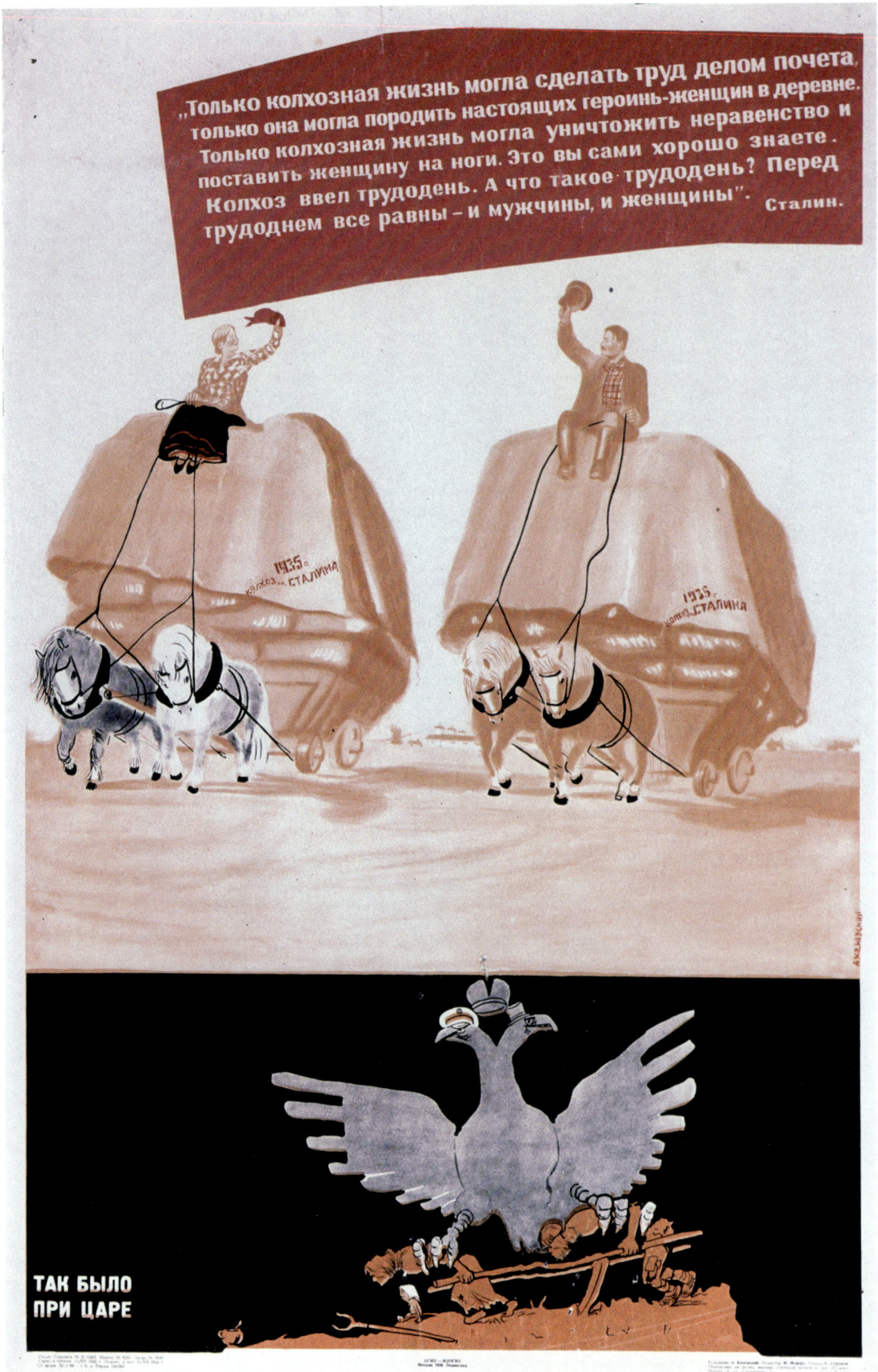
„Только колхозная жизнь могла сделать труд делом почета,
только она могла породить настоящих героинь-женщин в деревне.
Только колхозная жизнь могла уничтожить неравенство и
поставить женщину на ноги. Это вы сами хорошо знаете.
Колхоз ввел трудодень. А что такое трудодень? Перед
трудоднем все равны – и мужчины, и женщины".
Сталин.
1935 г. Колхоз им. Сталина
1935 г. Колхоз им. Сталина
ТАК БЫЛО
ПРИ ЦАРЕ

the course of one year collective farms had three to five different chairmen. Another publication mentioned one collective farm that had twelve different chairmen in 1934.[2] Party purges and political infighting were the main reasons for the rapid turnover. In 1934, ongoing purges also contributed to the free fall of the number of party cells in collective farms countrywide, from 30,000 to 18,300.[3] The country also continued to experience serious crop failures over the next couple of months, and the authorities continued to hunt down scapegoats. On May 28, the government admitted publicly that drought was threatening crops, especially in Ukraine. In Kiev, moreover, hundreds of suspects, including many Communists and senior government officials, had been arrested and were being arraigned on charges of "wrecking" the agricultural supply system, by "squandering, misappropriating, and disorganizing food supplies." Several show trials were being prepared. The malfeasance was, according to the Soviet press, discovered thanks to "direct intervention by Comrade Stalin."[4] Ironically, party officials also denounced government statisticians for overestimating the harvest of 1933, which of course they could have done only on orders from party officials.

Nor did the earnest and eager hard work on collective farms witnessed in 1933 continue. Stalin and Molotov issued a directive on August 31, 1934, obligating collective farms first to complete the planned grain delivery plan and then to create reserves for fulfilling planned grain purchases for the needs of the farm—before dividing up what was left over into "labor days." In many instances, this requirement significantly reduced wages. Consequently, the majority of collective farmers put in relatively few labor days. Absenteeism was rife and labor discipline low. Most collective farmers got into the habit of focusing their hard work on their individual plots of land and allowable individual livestock. It was certainly not uncommon, moreover, for collective farmers to use socialized livestock for their own purposes. They also typically treated the two kinds of animal differently. Those belonging to the collective were neglected, those of the household doted on.

The top-heavy hierarchy of officials who did not contribute directly to agricultural output also had to be fed from the common fund of grain. Their ranks included a collective farm chairman, one or more deputy chairmen, secret police officials, and bookkeepers. Tractor drivers, who of course contributed directly to production, also formed part of the rural elite and received the highest wages of any agricultural workers. Many wives of the collective farm elites did not toil, while other women often were not released from fieldwork until the eve of childbirth. The collective farm elite typically snatched up scarce consumer goods, like shoes, sponges, and matches. This was often also true of bread. The availability of locally produced goods had suffered from the expulsion of many village craftsmen during the anti-"kulak" campaign. Officially, to gain entry to the party and the Soviet administration, one needed a poor, proletarian background. In practice, however, a host of energetic adventurists were able to attach themselves to the new system, having first forged the necessary genealogies for themselves. The constant process of exposure and unmasking removed a portion of these people, but their places were immediately taken by others.

Stalin and the other party leaders were convinced that the relatively scanty wages were more than compensated by greater access to education and medical care, though presumably most collective farmers would have traded these services for economic independence, given the age-old peasant desire to be left alone.

Fig. 49 A.M. Kanevsky. "Only collective farm life has transformed labor into a point of honor and has made possible true heroines in the countryside." 1936. "Only collective farm life has transformed labor into a point of honor and has made possible true heroines in the countryside. . . . The collective farm has introduced the labor day. And what is the labor day? It has made all equal, men and women.—Stalin." Below: "As it was under the tsars."

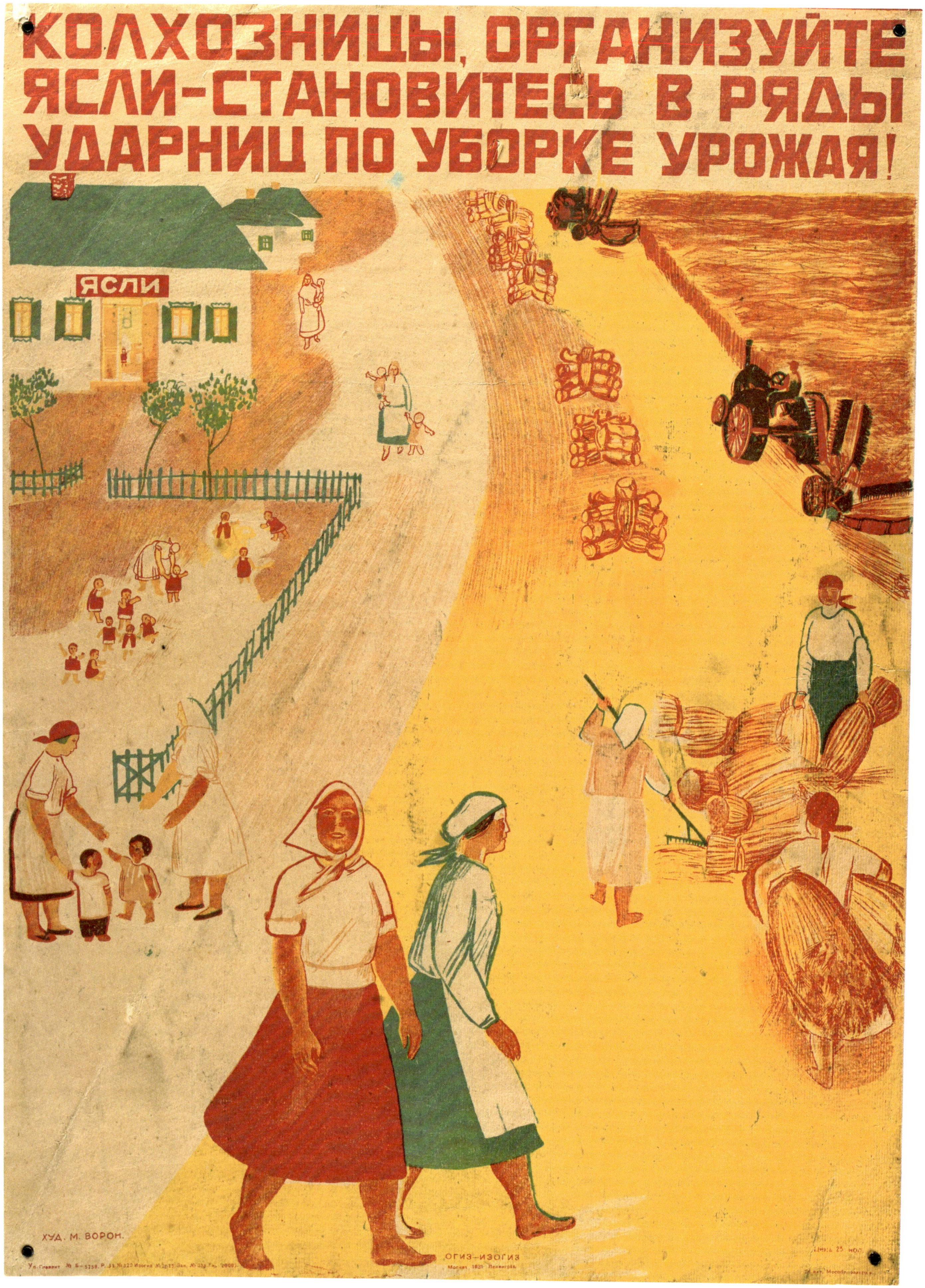
КОЛХОЗНИЦЫ, ОРГАНИЗУЙТЕ ЯСЛИ–СТАНОВИТЕСЬ В РЯДЫ УДАРНИЦ ПО УБОРКЕ УРОЖАЯ!
ЯСЛИ
ХУД. М. ВОРОН.
ОГИЗ–ИЗОГИЗ
Москва 1931 Ленинград
Цена 25 коп.

This tendency presumably contributed to collective farmers and the farm hierarchy banding together to resist encroachments from the outside world, with most farmers devoting the lion's share of their time and effort to their own household output and the elites defending them from outside pressure. Village solidarity remained stronger than the bonds of class or ideology, a reality that "could only negate the Communist dogma on class struggle and expose collectivization for what it was, a virtual Civil War between the state and the peasantry," in the words of the doyenne of Soviet collectivization studies in the English-speaking world, Lynne Viola.[5] Of course, it is not much of a civil war when one of the two sides is all but powerless and can resist only with pitchforks, foot dragging, the slaughter of livestock, and flight. Not a civil war but a decimation.

The bargain thus struck meant that the government consistently received a mammoth share of grain and meat output—30–40 percent—while labor productivity on the collective farms was significantly lower than that achieved by the more productive free peasants before collectivization. This outcome is what Kaganovich, on September 4, 1935, called "our absolutely unprecedented, stupefying victory, a victory of Stalinism."[6] In other words, the regime took control of the countryside but at the cost of generalized inefficiency. Nor did the Soviet population at large eat as well in the mid-1930s as it had in 1928. Only potatoes and vegetables were more abundant.

Even in the best of times, the regime could not let down its guard for fear of enemies. Hitler had come to power at the start of the previous year, and a more ardent anti-Communist and anti-Bolshevik was hard to conceive. On June 8, 1934, the government passed a law against treason—"betrayal of the Motherland [*Rodina*]"—with capital punishment as the recommended punishment. Actions defined as treasonous in most countries, including Russia before 1917, were called "counterrevolutionary" in Soviet Russia. The Revolution, in other words, was held to be the highest collective value. Such terminology was henceforth retained in the criminal code, but the state itself was now represented as equally important. Similarly, on July 10 the secret police was reorganized and made into a People's Commissariat for Internal Affairs (or ministry), the NKVD, with Iagoda in charge. Such were hallmarks of what Nicholas Timasheff called the "Great Retreat" from revolutionary ideals and a return to traditional values.[7] In reality, nearly all of the characteristics of the revolutionary Communist regime were retained, including the abolition of private property and the free market, the collectivized agricultural sector, central economic planning, tight party control of culture, militant atheism, strict state censorship of all media, one-party dictatorship, pervasive secret police vigilance, and subversive political operations worldwide, among others.

The June 8 law, moreover, incorporated an element entirely absent from traditional conceptions of treason, namely, the concept of the collective responsibility of family members. Those related to a traitor were to expect harsh punishment as well. This interpretation of the law surely derived from the Marxist class interpretation of guilt, which had been a cornerstone of Soviet jurisprudence from immediately after the revolution and had influenced the punishment of family members of "kulaks." Henceforth, guilt by association with traitors and counterrevolutionaries would be a routine feature of Soviet criminal justice and politics.

Even the apparent subordination of the secret police to the state did not in any way diminish its scope, ubiquity,

Fig. 50 M.A. Voron. "Collective farm women, organize daycare and join the ranks of female shock workers in gathering the harvest!" 1931.

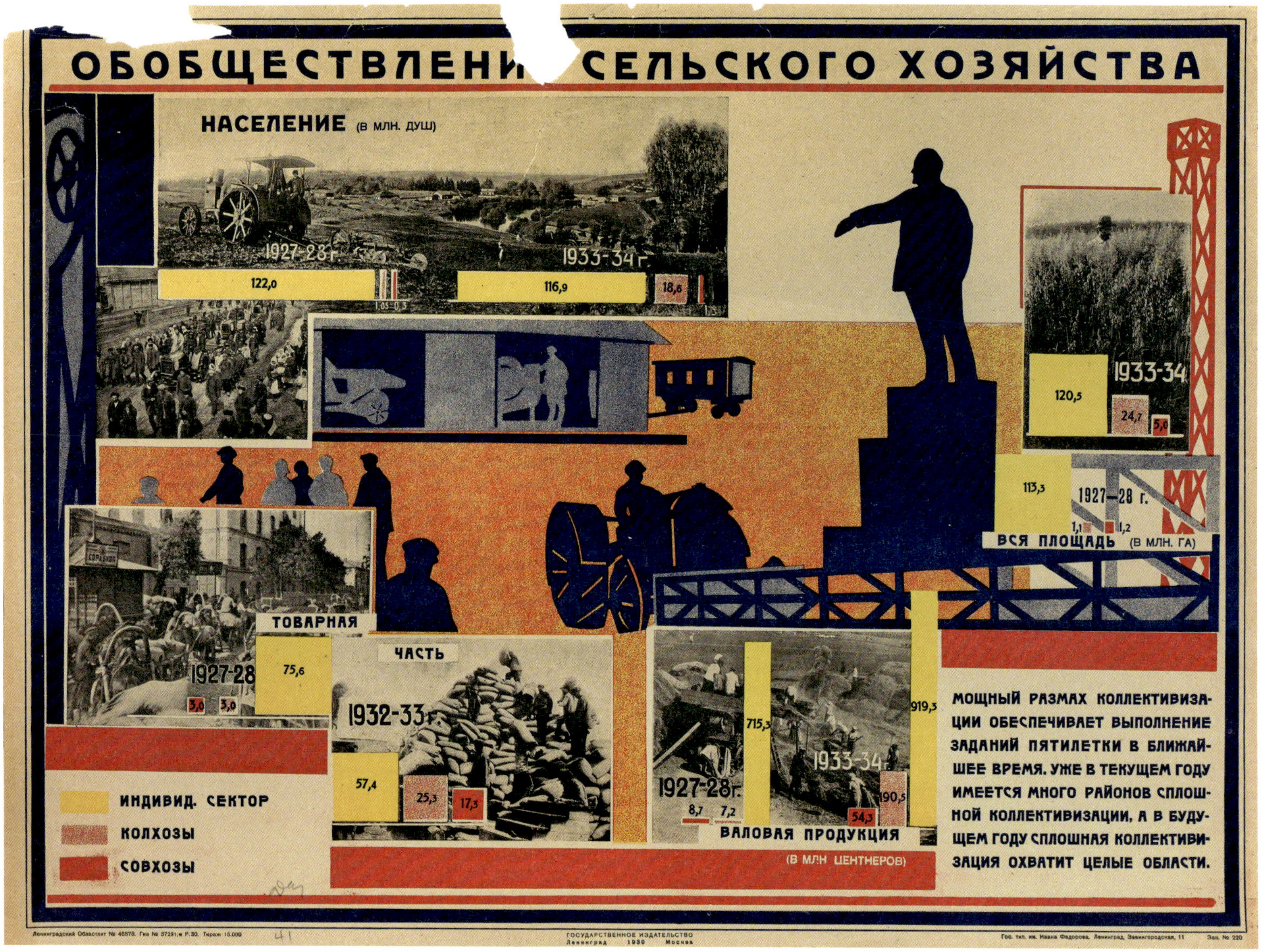

Fig. 51 Artist not identified. "The socialization of agriculture." 1930. "The powerful advancement of collectivization will secure the fulfillment of the goals of the Five-Year Plan in the near future . . ."

power, and extrajudicial authority. When on December 1, the popular Leningrad party boss, Sergei Kirov (1886–1934), was murdered under deeply mysterious circumstances, the NKVD acted with no more inhibition than the OGPU and in fact with much less, thanks to a law drafted by Stalin and adopted that very night to simplify legal procedure. The law required "terrorism" cases to be investigated and tried immediately, and death sentences to be carried out upon conviction. Within days, seventy people had been tried for "preparing terrorist actions against Soviet authorities"; sixty-six were sentenced to death.[8] Thus the groundwork was laid for the infamous Moscow show trials of 1936–38, in which thousands of party members—and tens of thousands of "kulaks"—were executed.

Meanwhile, material conditions in Soviet Russia somewhat improved. On January 1, 1935, the rationing of flour and bread ceased, though prices increased sharply.[9] New rules adopted on February 17 at the Second Congress of Collective Farm Shock Workers allowed collective farmers access to individual garden plots of 0.25–0.5 hectare (0.6–1.2 acres)—up from 0.15 hectare—and possession of one cow, two calves, one pig, ten sheep, and unlimited fowl.[10] The statute also permitted collective farmers to sell their produce at collective farm markets and to make use of socialized workhorses in their own gardens for a fee. To a large extent, the new rules merely conferred official sanction on practices that had become routine. In any event, at the start of 1936, only 23.3 percent of collective farmers in the USSR had neither cow nor calf nor heifer.[11] For women this was especially important, because they played a big role in livestock production.

The collective farm woman labored under a crushing double burden. Using mostly primitive tools and utensils, she cooked, baked, sewed, wove, washed, and hauled water. She also tended the family's garden and livestock, as well as her children. Less than 10 percent did no further work outside the home. In fact, over 78 percent accomplished thirty-five or more days each year of work on the collective farm.[12] The Ukraine-born economist Naum Jasny (1883–1967) marveled that a Soviet agricultural journal, while stressing that it was "particularly significant to assure an adequate rest to the high-yielding cows, which are characterized by a considerable nervous excitability,"[13] only mentioned in passing the need for rest of the women who tended them.

A milestone was reached in the history of Soviet labor on August 31, 1935, when a coal miner in the eastern Ukrainian Donbas region, Aleksei Stakhanov (1906–77), apparently cut 102 tons of coal, or fourteen times his normal quota.[14] Though a number of circumstances made this feat possible, the Soviet authorities rushed to hold him up as a role model for workers throughout the economy. Prizes and incentives of all kinds were offered to "Stakhanovites" who overfulfilled their work quotas, in both industry and agriculture. At a conference of Stakhanovites that met in the Kremlin on November 17, 1935, Stalin boasted that the Soviet Union had achieved so much progress that "life has improved, comrades. Life has become more joyous. And when life is joyous, work goes well. Hence the high rates of output."[15] His words, uttered in the middle of what became known as the "three good years," were picked up by propagandists and became pervasive in Soviet society. And indeed, by October 1, the rationing of meat, oil, fish, sugar, and potatoes had ceased, though again prices sharply increased.

In his report to the Eighteenth Party Congress on March 10, 1939, Stalin declared that Soviet agriculture "is not only run on the largest scale, and is the most mechanized in the world, and therefore produces

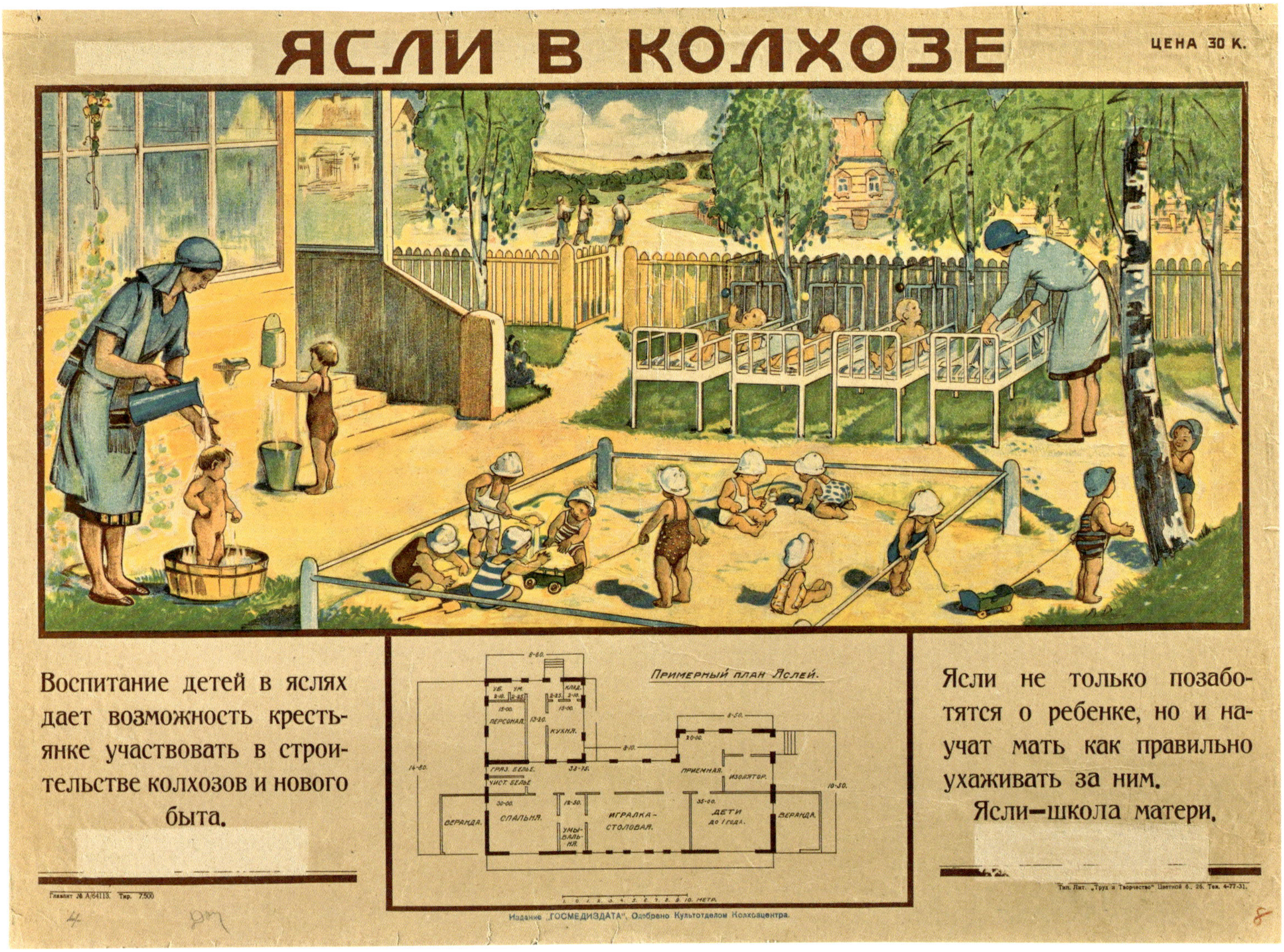

Fig. 52 Artist not identified. Moscow: "Daycare in the collective farm." 1917–30. "Children going to daycare give collective farm women the opportunity to take part in building the collective farm and the new way of life."

the largest surplus for the market, but is also more fully equipped with modern machinery than the agriculture of any other country."[16] For the most part, this statement was wildly inaccurate. Total horsepower in the USSR had declined from roughly 29 million in 1928 to roughly 28 million in 1939.[17] In that year, significantly more horsepower was produced by horses than by tractors. Oftentimes, moreover, after only a few years of operation, tractors were ready for the scrap heap, though they were still registered with the MTS. A sharp decline in new tractors from 1937 meant that more repairs were necessary. Insufficient motor fuel and the frequent departure of skilled tractor drivers in search of better opportunities resulted in long spans of time when even functional tractors were not running. Weak mechanization and low draft power naturally depressed crop yields. For grain, the trend was only slightly upward, from 7.5 centners per hectare in 1928–32 to 7.7 in 1938–40.[18] Nor did grain exports return to the highs of the early 1930s; in 1938, only 2 million tons were exported, thanks not to great efficiency but to extremely big procurements. (By contrast, in 1913, Russia had exported around 10 million tons.) Livestock exports all but ceased by the end of the decade.

Furthermore, in 1938, individual collective farmers, eking out their harvests with the aid of a horse or ox and all hands in the household wielding primitive tools, produced 52.1 percent of potatoes and vegetables, 56.6 percent of fruit, 31.4 percent of milk, 70.9 percent of meat, and 70.4 percent of leather—for the entire country.[19] They had to deliver roughly 20 percent of the milk and potatoes they produced and some 30 percent of the meat to the government as a form of tax. The rest they could sell in the collective farm markets. Of course, since the farmers had to haul the produce themselves, sometimes over fairly long distances, a task they were rarely well equipped to perform, much labor was wasted.

The USSR was obviously not a highly mechanized country with an efficient agriculture. In fact, a few years after Stalin's death, the country would begin to import grain and, from 1963, big quantities of it (nearly 10 million tons in that year).[20] With the order of May 27, 1939, "for the defense of collective farm property from dissipation,"[21] the government moved to shrink the individual plots, a policy it hesitantly pursued for the next two years but backed away from once the war began, given the serious hardship it had caused. Thus, the collective farmers working on their individual plots and tending their little herds of livestock would continue to contribute 30–50 percent of nongrain agricultural products in the Soviet Union until the fall of Communism.

In the collective fields, they worked far less. In 1937, more than 10 percent of all collective farmers did not earn one single labor day. As an official in Saratov put it, "Collective farmers do not receive much bread or money for their labor days, so they do not work hard."[22] At the May plenum of 1939, therefore, the party leadership decided to require collective farmers to complete one hundred labor days (or sixty to eighty in cotton-producing regions).[23] Collective farm directors were given significant powers to enforce these regulations—for example, to take away individual plots.

Among those least willing to work hard were village young people, among whom, ironically, the new rulers found their principal mass support. After all, they had grown up in the Soviet era and had gone to Soviet schools. They had not absorbed traditional village culture's deep respect for stalwart labor. But they had learned a great deal, in school, about the poverty and unhappiness of village

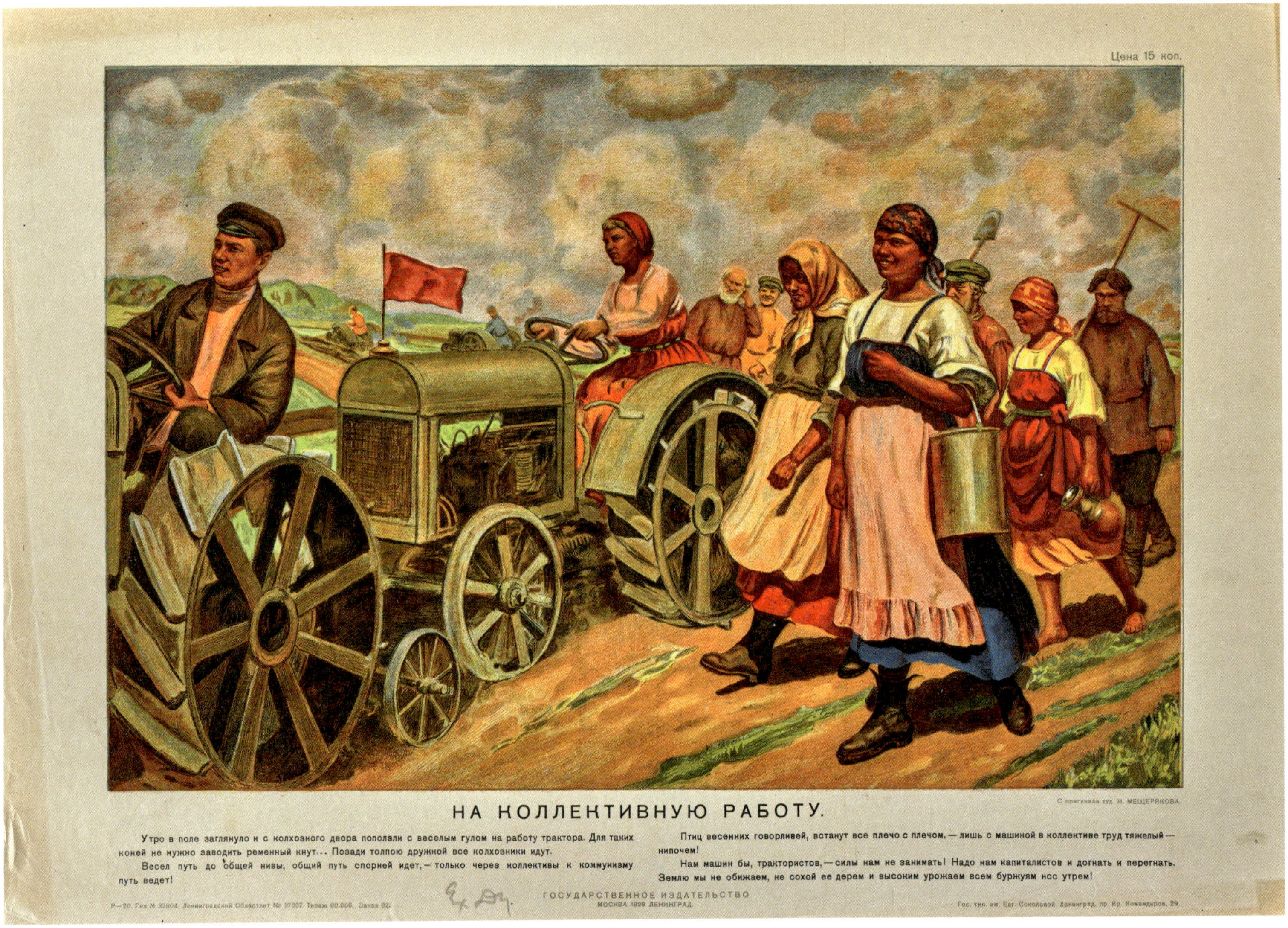

▲ **Fig. 53** I. Meshcheriakov. "To the collective work." 1929. "In the morning, one glances into the field and there tractors are setting out from the collective farm with a cheerful hum . . ."

▶ **Fig. 54** V.A. Serov. "Widen the front of collective farm trade." 1932. "Decree of the Central Committee and the Council of People's Commissars of the USSR on the rules governing collective farm trade."

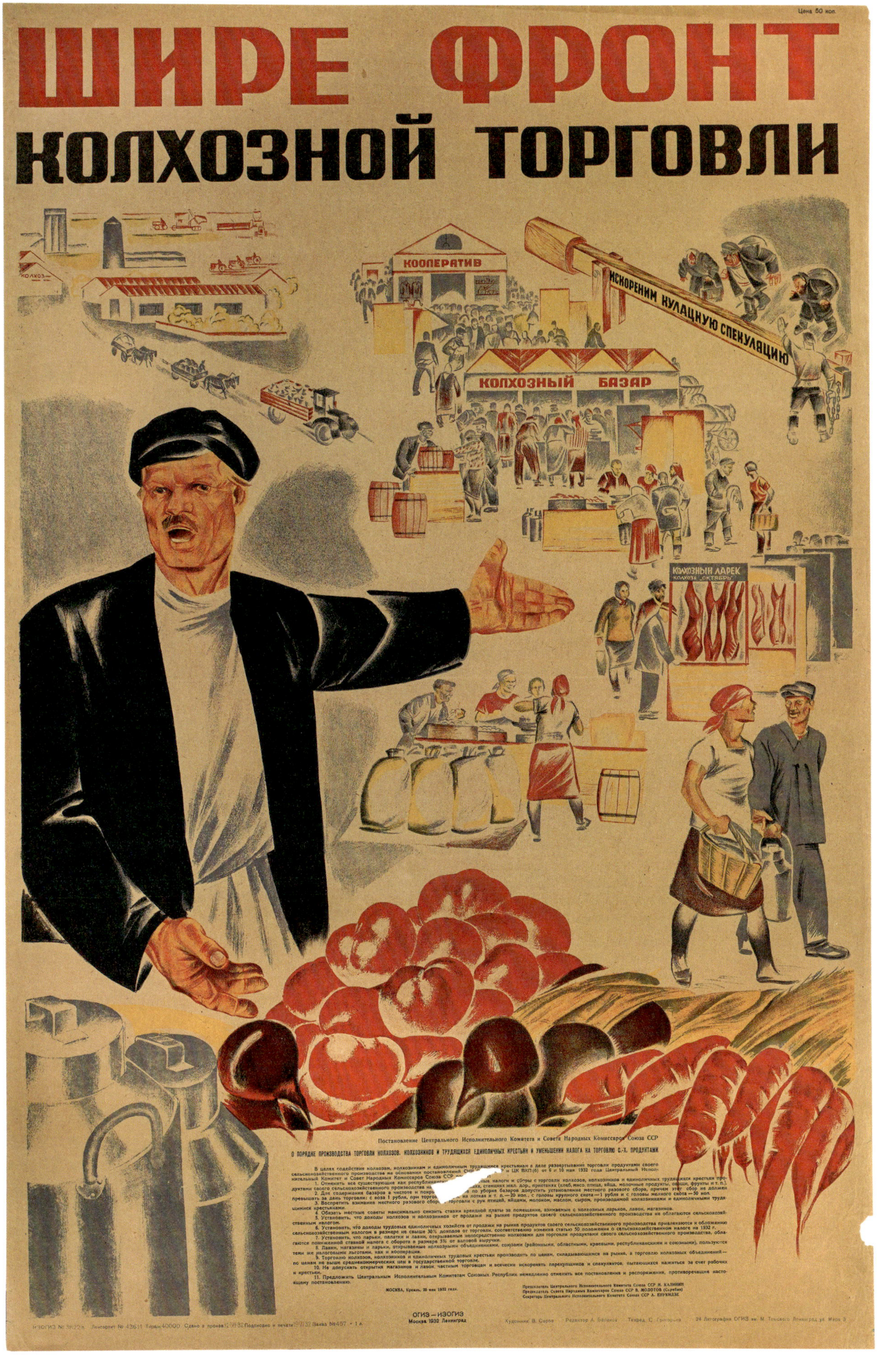
Цена 50 коп.
ШИРЕ ФРОНТ
КОЛХОЗНОЙ ТОРГОВЛИ
КООПЕРАТИВ
ИСКОРЕНИМ КУЛАЦКУЮ СПЕКУЛЯЦИЮ
КОЛХОЗНЫЙ БАЗАР
КОЛХОЗНЫЙ ЛАРЕК
КОЛХОЗ „ОКТЯБРЬ"
Постановление Центрального Исполнительного Комитета и Совета Народных Комиссаров Союза ССР
О ПОРЯДКЕ ПРОИЗВОДСТВА ТОРГОВЛИ КОЛХОЗОВ, КОЛХОЗНИКОВ И ТРУДЯЩИХСЯ ЕДИНОЛИЧНЫХ КРЕСТЬЯН И УМЕНЬШЕНИИ НАЛОГА НА ТОРГОВЛЮ С.-Х. ПРОДУКТАМИ
В целях содействия колхозам, колхозникам и единоличным трудящимся крестьянам в деле развертывания торговли продуктами своего сельскохозяйственного производства на основании постановлений СНК ... Р и ЦК ВКП(б) от 6 и 10 мая 1932 года Центральный Исполнительный Комитет и Совет Народных Комиссаров Союза ССР ...
1. Отменить все существующие как республикан... ные налоги и сборы с торговли колхозов, колхозников и единоличных трудящихся крестьян продуктами своего сельскохозяйственного производства на ... станциях жел. дор., пристанях (хлеб, мясо, птица, яйца, молочные продукты, овощи, фрукты и т. п.).
2. Для содержания базаров в чистоте и покр... по уборке базаров допустить установление местного разового сбора, причем этот сбор не должен превышать за день торговли: с воза 1 рубля, при торгов... на лотках и т. п. — 20 коп., с головы крупного скота — 1 рубля и с головы мелкого скота — 50 коп.
3. Воспретить взимание местного разового сбора ... торговли с рук птицей, яйцами, молоком, маслом, сыром, производимой колхозниками и единоличными трудящимися крестьянами.
4. Обязать местные советы максимально снизить ставки арендной платы за помещение, взимаемые с колхозных ларьков, лавок, магазинов.
5. Установить, что доходы колхозов и колхозников от продажи на рынке продуктов своего сельскохозяйственного производства не облагаются сельскохозяйственным налогом.
6. Установить, что доходы трудовых единоличных хозяйств от продажи на рынке продуктов своего сельскохозяйственного производства привлекаются к обложению сельскохозяйственным налогом в размере не свыше 30% доходов от торговли, соответственно изменив статью 50 положения о сельскохозяйственном налоге на 1932 г.
7. Установить, что ларьки, палатки и лавки, открываемые непосредственно колхозами для торговли продуктами своего сельскохозяйственного производства, облагаются пониженной ставкой налога с оборота в размере 3% от валовой выручки.
8. Лавки, магазины и ларьки, открываемые колхозными объединениями, союзами (районными, областными, краевыми, республиканскими и союзными), пользуются теми же налоговыми льготами, как и кооперация.
9. Торговлю колхозов, колхозников и единоличных трудовых крестьян производить по ценам, складывающимся на рынке, а торговлю колхозных объединений — по ценам не выше среднекоммерческих цен в государственной торговле.
10. Не допускать открытия магазинов и лавок частным торговцам и всячески искоренять перекупщиков и спекулянтов, пытающихся нажиться за счет рабочих и крестьян.
11. Предложить Центральным Исполнительным Комитетам Союзных Республик немедленно отменить все постановления и распоряжения, противоречащие настоящему постановлению.
МОСКВА, Кремль, 20 мая 1932 года.
Председатель Центрального Исполнительного Комитета Союза ССР М. КАЛИНИН
Председатель Совета Народных Комиссаров Союза ССР В. МОЛОТОВ (Скрябин)
Секретарь Центрального Исполнительного Комитета Союза ССР А. ЕНУКИДЗЕ
ОГИЗ — ИЗОГИЗ
Москва 1932 Ленинград
Художник В. Серов

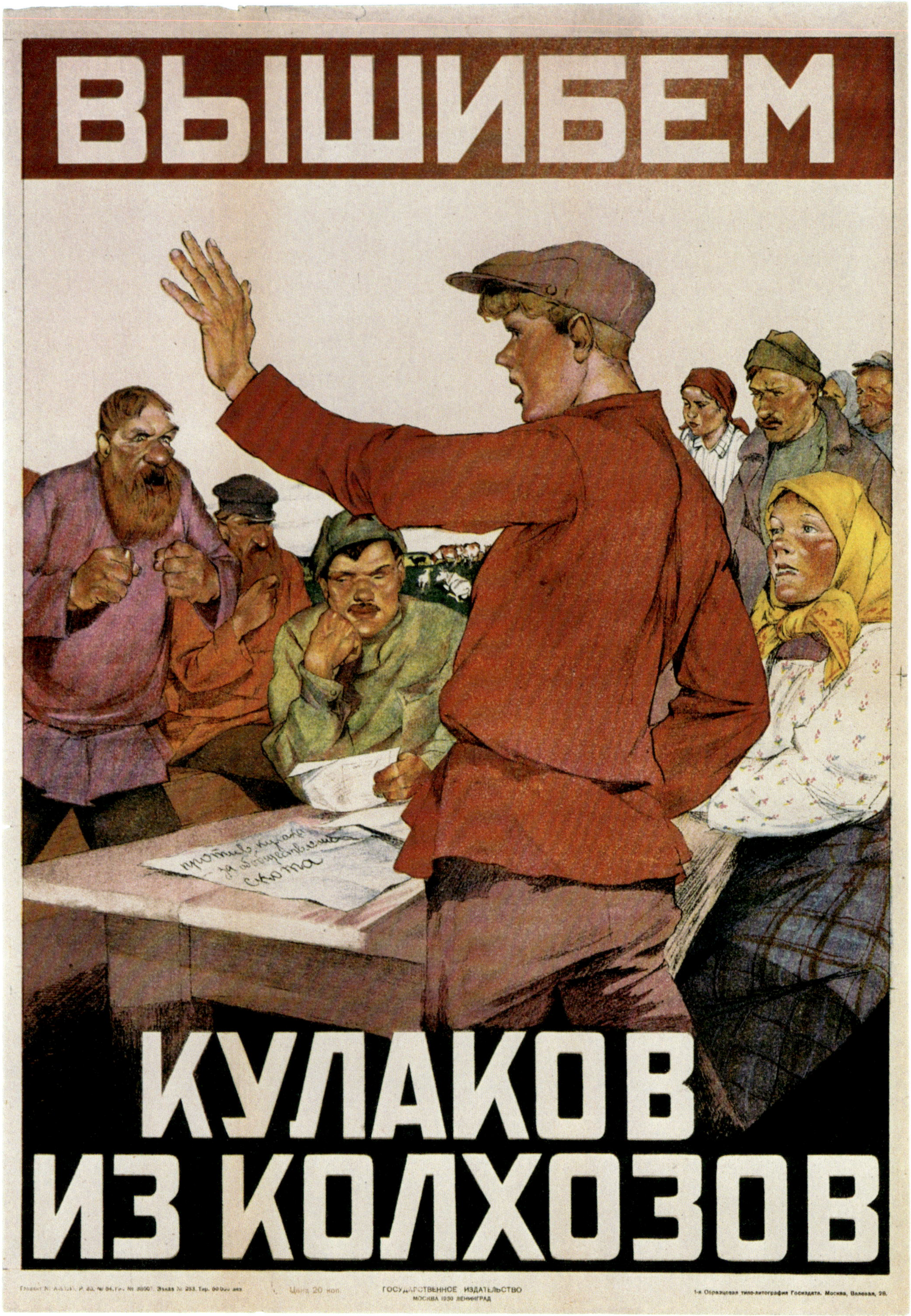
ВЫШИБЕМ
КУЛАКОВ
ИЗ КОЛХОЗОВ
Цена 20 коп.
ГОСУДАРСТВЕННОЕ ИЗДАТЕЛЬСТВО
МОСКВА 1930 ЛЕНИНГРАД
1-я Образцовая типо-литография Госиздата. Москва, Валовая, 28.

life, which corresponded to their own observations. Change was the watchword of the time: the need to establish a new, better, more just order. The young people believed it was up to them to carry out these necessary changes for the better. They rushed to the cities to study and, if they remained in the countryside, were in no hurry to perform arduous and disrespected agricultural work. The opponents of the new and better life, in the minds of the youth, were the representatives of the older generation, whose conservatism, instinct for private property and capital accumulation, ignorance, and backwardness were a hindrance to the coming of a better tomorrow in which labor would be easy, consumption high, and life joyful.

The leadership used a variety of incentives to stimulate hard work, such as sanitarium visits, bonuses in kind, and vacations. The best producers, called "shock workers" (*udarniki*), could also be sent to regional and national meetings. The young men and women were especially interested in achieving shock worker status because they were excluded from other means of advancement. Encouragement of shock work had begun on a small scale in 1930 and had gradually gained importance. A first All-Union Congress of Collective Farm Shock Workers had met in Moscow in February 1933, with 1,513 delegates attending. Most were registered as poor (56.8 percent), older than thirty (62.5 percent), and male (85.2 percent). Whereas the average collective farmer in that year had accumulated 118 labor days, 73 percent of the delegates had more than 257, and 19 percent over 400.[24] At the second congress, in November 1935, for the most part only ordinary peasants presented speeches, though over half of them were party members and local officials.

Of course, it was a big deal for all of them to attend a meeting with Stalin present, since few had ever traveled farther than the next provincial town. Maria Demchenko, along with thirty-one other sugar-beet heroes, only one of them a man, received the Order of Lenin. Praskovia (Pasha) Angelina (1912–59), one of the first female tractor operators in the USSR, was also honored. She had a personal conversation with Stalin himself and addressed the Congress. She went on to a meteoric career. In 1937, she joined the party and was elected to the Supreme Soviet. She took part in the eighteenth and subsequent party congresses, beginning in 1939, and received many titles and honors, including "Hero of Socialist Labor" twice.[25] Streets and a steamship were named after her. Stalin and the party leaders did not view such direct personal involvement with the details of agricultural production as a waste of time. In fact, from 1930 to 1936, they carefully monitored grain production and collection "region-by-region, month-by-month, and even every five days," according to an eminent specialist, R. W. Davies.[26]

Not everyone was as happy as Stalin with the shock workers' successes. As with the Stakhanovite movement, resentment festered among those who missed out on the honors. The feats of shock work, moreover, often required a lot of support from fellow workers. Stakhanov himself could not have hewn those 102 tons, for example, without proppers shoring up the tunnel as he dug into it. Yet they received no recognition. Such resentment led to sabotage. One woman who had received a prize for being the sixteenth-best milkmaid in the country returned home to find that her most cherished cow had died from ingesting fodder laced with sewing needles. In a similar case, a collective farm chairman refused to give a prize-winning milkmaid fodder for her cows. "Any idiot can increase milk output with fodder," he said. "Let's see you increase it without."[27]

Fig. 55 Artist not identified. "We will kick the kulaks out of the collective farms." 1930.

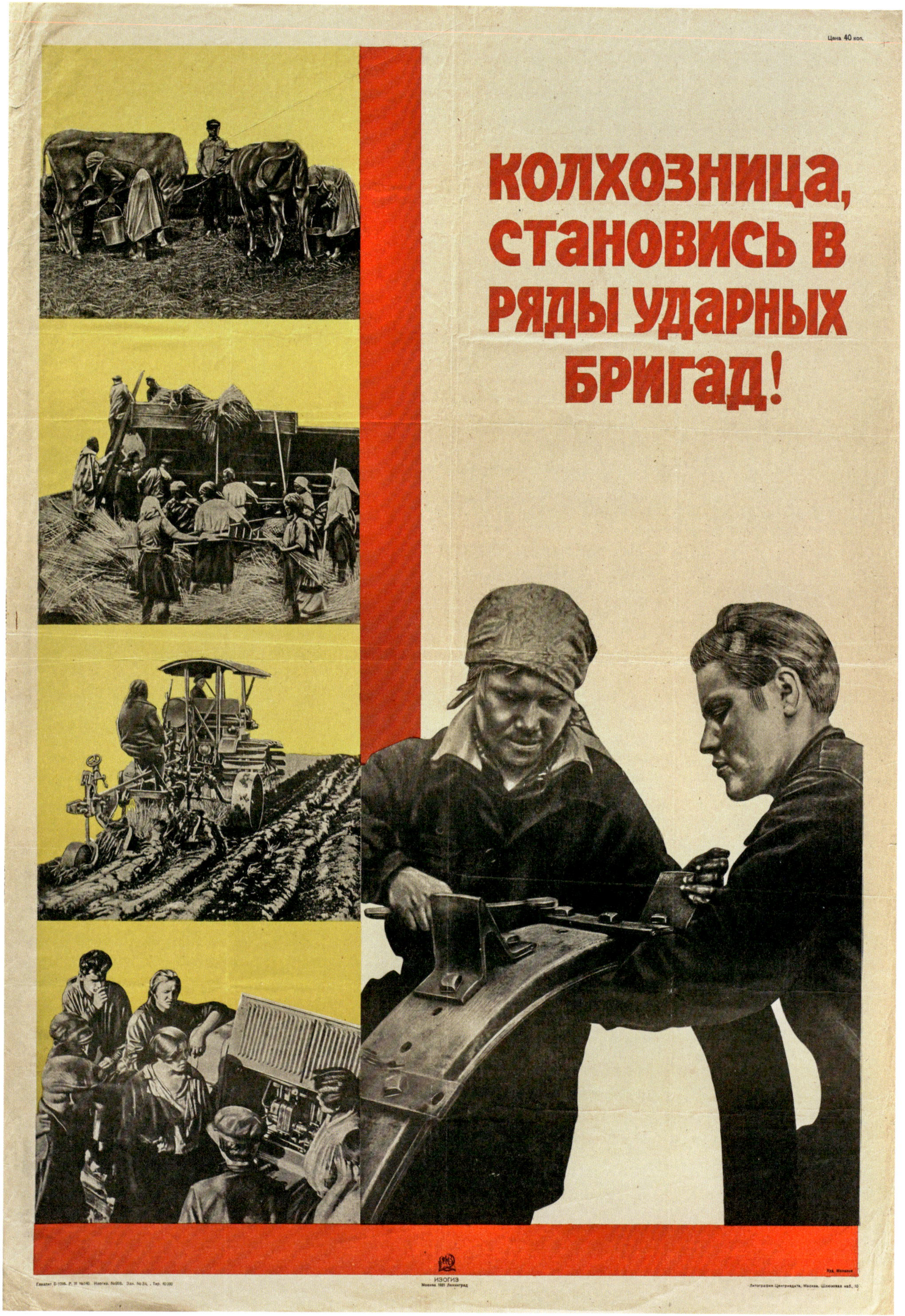
Цена 40 коп.
КОЛХОЗНИЦА,
СТАНОВИСЬ В
РЯДЫ УДАРНЫХ
БРИГАД!
ИЗОГИЗ
Москва 1931 Ленинград
Литография Центриздата, Москва, Шлюзовая наб., 10

Such instances of sabotage were fairly frequent and involved everything from damage to crops and livestock to humiliation and even murder. Local officials were often reluctant to prosecute perpetrators, despite frequent centrally promoted campaigns aimed at rooting them out.

In the late 1930s, collectivization continued to encompass a larger share of agricultural producers, but more through attrition of older individual farmers than through active campaigning. The total number of peasant households in the country declined from 25.2 million in mid-1928 to 20.1 million in July 1938. The decline was due to the migration of several million people from the countryside to the cities, seeking education and work. At this point, only 1.31 million individual farmers remained.[28] Of course, working as an independent farmer without government officials breathing down one's neck was the ideal for nearly all Soviet peasants. It was also, ironically, the more efficient and effective arrangement of agricultural production on average. After the good harvest of 1937, the number of individual farmers increased somewhat, while fewer joined collective farms. Individual farmers could do quite well renting out their services as haulers. But a new tax on the private ownership of horses instituted in fall 1938 closed the last loophole of independent existence outside of state and collective agriculture.[29] By this point, a large proportion of individual farmers were simply old people who had escaped the anti-"kulak" campaigns. The almost complete elimination of individual farmers was brought about largely by the law of May 27, 1939, mentioned above, which reduced the amount of farmland individual farmers could have to 1 hectare.

So what happened to the "kulaks"? A decree of the Central Executive Committee of January 25, 1935, restored voting rights to all former "kulaks,"[30] and the 1936 "Stalin" Constitution reaffirmed their rehabilitation. However, as with so many other guarantees enshrined in that document—such as freedom of speech, assembly, and religion, which never gained legal force—few of those still in exile actually benefited. In fact, many former "kulaks" actually suffered: by registering to have their rights restored, they only made themselves more visible to the police, with baleful results. By 1936, all the children of special settlers were in school, but they often studied in two shifts because of inadequate facilities.[31] Thousands were liberated to study, having married people in other social categories, or for other reasons. But these changes did not significantly affect "kulak exile."

The most terrifying events for the "kulaks" and indeed for the entire country in these years was the Great Terror, when the NKVD conducted several "mass operations." On July 2, 1937, the Politburo, at Stalin's instigation, directed all regional party organizations to compile lists of "former kulaks, criminals, and other anti-Soviet elements."[32] Stalin was deeply fearful that such people had snaked their way into Soviet institutions, especially collective farms, the railroads, and factories. Officials throughout the country were also frightened by the millions of people constantly wandering, moving from job to job, living in temporary circumstances, not fitting into official categories, including the unemployed, orphans, "hooligans," and national minorities. Ironically, it was precisely the collectivization and "dekulakization" drives that had uprooted so many millions and cast them into uncontrolled movement, searching for (often suspicious) ways to fit in. On July 30, Nikolai Ezhov (1895–1940), the head of the NKVD who had replaced Iagoda in 1936, signed Operation Order No. 00447.[33] It became known as the "kulak operation,"

Fig. 56 Makhalov. "Collective farm woman, join the ranks of the shock work brigades." 1931.

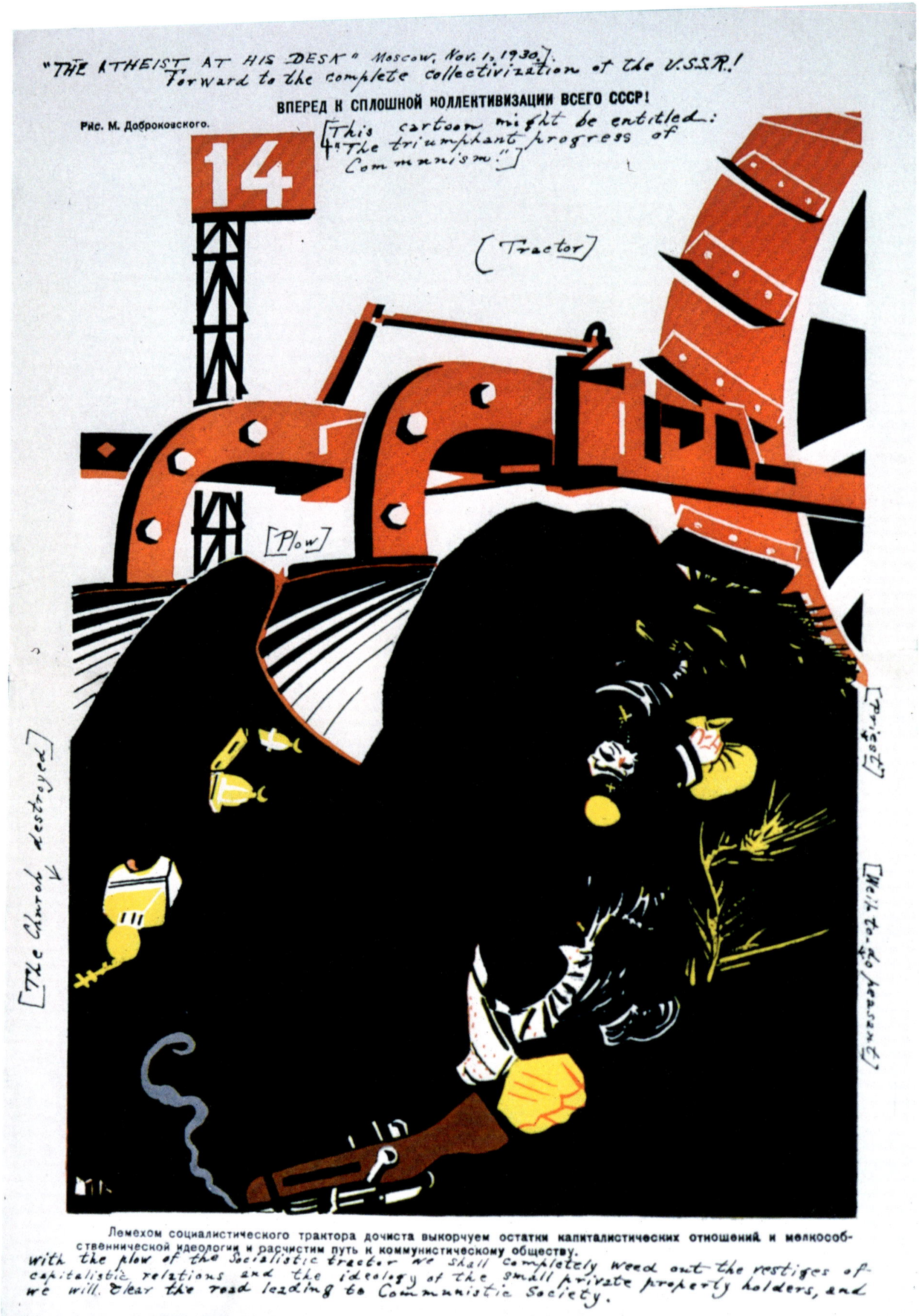

◀ **Fig. 57** M. Dobrokovsky. "Forward to universal collectivization of the USSR." 1929–32? "With the plowshare of the socialist tractor, we will completely root out the vestiges of capitalist economic relations and the petty-bourgeois ideology and will clear a path toward communist society."

▶ **Fig. 58** M. Soloviev. "The harvest to the front." Okno TASS Nº 1043. 1944. "Collective farmer! You must complete your work tasks with military precision. Collect the harvest, losing not a single grain, and fulfill the planned grain delivery ahead of time!" Poem by A. Mashistov.

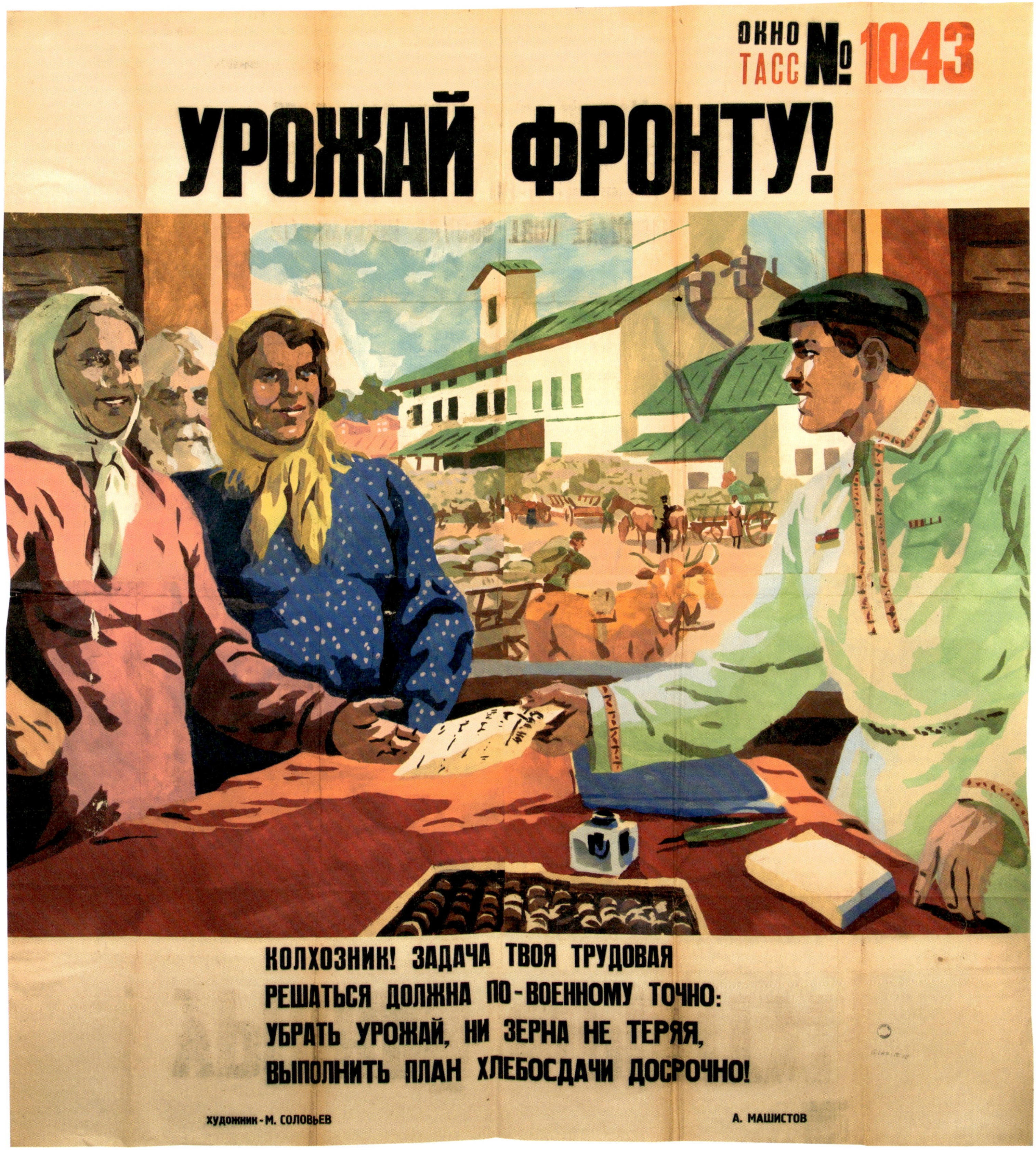
ОКНО ТАСС № 1043
УРОЖАЙ ФРОНТУ!
КОЛХОЗНИК! ЗАДАЧА ТВОЯ ТРУДОВАЯ
РЕШАТЬСЯ ДОЛЖНА ПО-ВОЕННОМУ ТОЧНО:
УБРАТЬ УРОЖАЙ, НИ ЗЕРНА НЕ ТЕРЯЯ,
ВЫПОЛНИТЬ ПЛАН ХЛЕБОСДАЧИ ДОСРОЧНО!
ХУДОЖНИК-М. СОЛОВЬЕВ
А. МАШИСТОВ

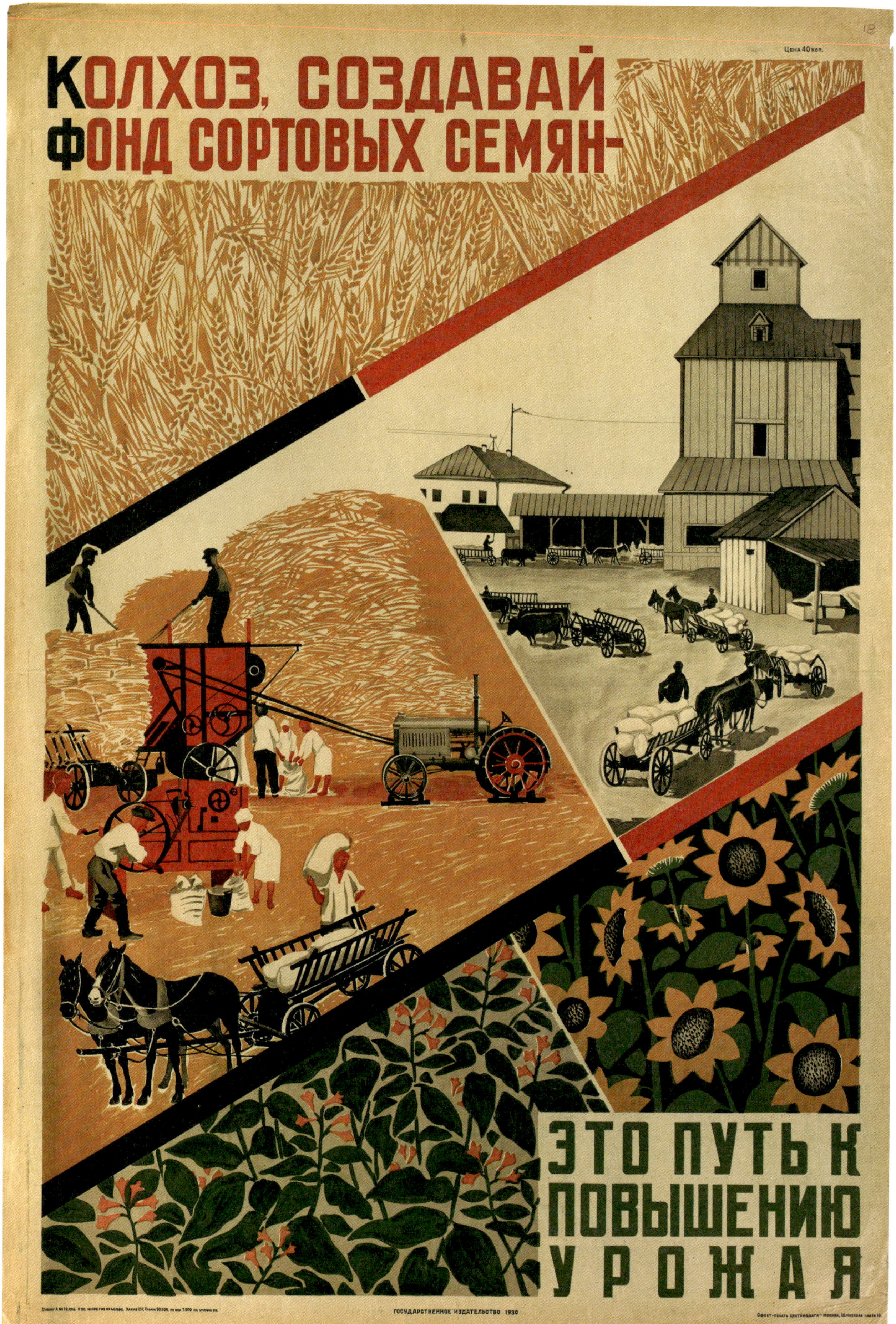
Цена 40 коп.
КОЛХОЗ, СОЗДАВАЙ
ФОНД СОРТОВЫХ СЕМЯН–
ЭТО ПУТЬ К
ПОВЫШЕНИЮ
УРОЖАЯ
ГОСУДАРСТВЕННОЕ ИЗДАТЕЛЬСТВО 1930
Офсет-печать Центриздата — Москва, Шлюзовая набер. 10.

Fig. 59 Artist not identified. "Collective farm, build up a fund of high-yield seeds. This is the path to an increased harvest." 1930.

since it affected them first and foremost. Yezhov added to Stalin's list "church people [*tserkovniki*]." He proposed that officials execute 70,000 and exile 186,500 of these unfortunates.[34] Out of fear of appearing too lenient, nearly all relevant authorities went beyond the required minimum. The result was the summary execution in 1937 and 1938 of, at the very least, 681,692 people.[35] It is unclear how many were actually former "kulaks." The antichurch campaign, the third major antireligious assault (along with those in 1922 and 1930), also resulted in the death and exile of many thousands of priests and the closing of thousands more churches.

A decree of October 22, 1938, gave sixteen-year-olds in "kulak" exile the right to obtain passports if they had no black marks against them.[36] Yet it had no immediate effect, because local officials were uncertain how to proceed. During the 1930s, roughly 230,000 babies were born in the settlements.[37] Several hundred thousand special settlers fled from exile during the 1930s. Of these, many must have perished in the forest or swamps, and 235,120 were caught and sent back.[38] As of July 1, 1938, according to official statistics, there were 997,329 special settlers living in 1,741 settlements,[39] nearly all victims of the anti-"kulak" campaign in 1929–33 (a minority had been sent as alleged saboteurs and other "anti-Soviet elements"). Most were being employed as forced labor throughout Soviet industry. During World War II, a huge number of collective farms fell under the control of the Nazis, who destroyed much infrastructure. In 1946, local authorities began requesting permission to reintegrate exiles into normal life, and these requests were generally granted. On August 13, 1954, a year and a half after Stalin's death, the Presidium of the Central Committee lifted all restrictions on the former "kulaks," though few could make up for lost time or live down the stain with which "dekulakization" had marked them. Two years later, in his Secret Speech, moreover, Nikita Khrushchev (1894–1971) did not mention the "kulaks" at all, as if that particular crime of Stalinism had not occurred.[40]

The collective farm system continued to limp along until the fall of Communism. The rural population gradually shrank, as anyone with ambition and talent strived to get out and into city life with its vastly greater amenities and comforts. The remaining farmers continued to produce insufficient grain to prevent hefty imports but also contributed 30 percent or more of the country's essential produce—including milk, eggs, meat, vegetables, and potatoes—on roughly 3 percent of the country's arable land during the last decades of the USSR. Many villages all but emptied out over the years or, if close to urban areas, filled up with vacationing heirs of deceased farmers. After the collapse of the Soviet system in 1991, the collective farms gradually disbanded.

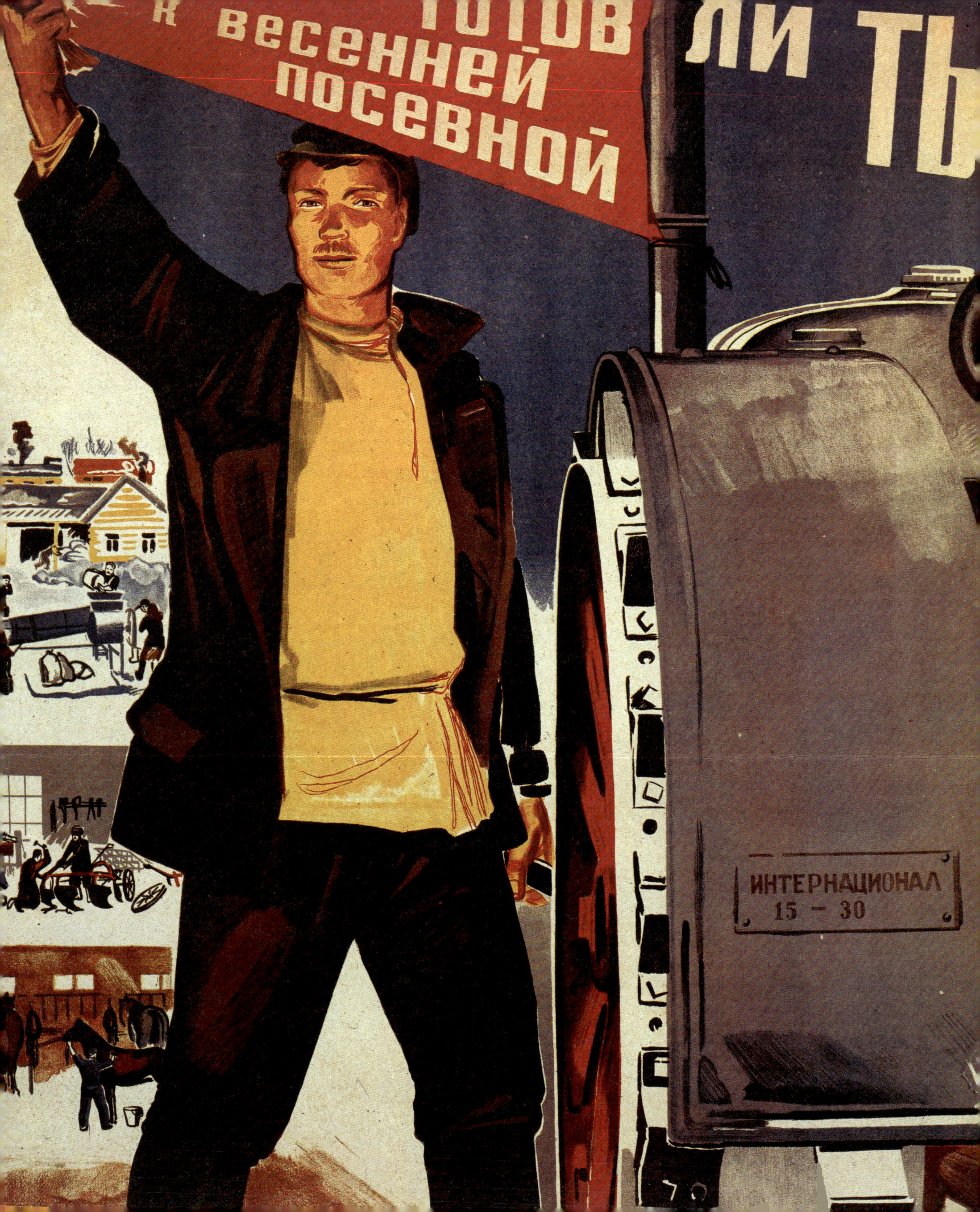
ли
к весенней
посевной
ИНТЕРНАЦИОНАЛ
15 – 30

CONCLUSION

Collectivization and "dekulakization" removed from the Soviet countryside a large proportion of competent and hardworking farmers. Given the obligation to follow centrally determined instructions, those remaining in the collective farms gradually forgot how to be peasants and how to farm with individual initiative. Thus, when Mikhail Gorbachev (b. 1931) came to power in 1985 and proposed the renovation of Soviet agriculture, he could not draw upon eager peasants chafing to reengage in commercial farming—unlike China under Deng Xiaoping. Equally important, the demand that all villagers put the interests of the collective and of society ahead of the interests of the individual and of the family produced the opposite result: a profound lack of civic-mindedness, an almost complete disengagement from public life, an intense focus on one's own individual garden and livestock, and the relative neglect of socialist property and tasks. The villages, moreover, continued to lose their most dynamic members, who migrated to the cities in search of better opportunities. Those who stayed behind suffered dramatically high levels of alcoholism. In other words, collectivization was a disaster for the USSR, which from the late 1950s systematically imported grain from the West.

Collectivization was obviously a thoroughgoing continuation of Bolshevik Civil War policies, such as the forced confiscation of grain, criminalization of market relations, dispatch of armed bands of Communist true believers into the countryside to take control, and incitement of hostility between social classes. There was even a brief but deadly episode at the start of 1919, when the Bolshevik leadership determined that the only way to defeat the Cossacks was through a "merciless struggle against the entire Cossack elite by means of their total extirpation."[1] Within the span of two months, ten or twelve thousand Cossacks had been massacred, after which the party backed off and Cossacks were no longer specifically targeted. "Dekulakization" was essentially the same policy—rooting out the elite to dominate the masses—though applied on a vastly bigger scale and without retreat. Similarly, the policies of violent grain confiscation and execution of grain peddlers were set aside with the New Economic Policy (NEP) in early 1921, after it had become obvious that the civil war economic policies (war communism) had devastated the economy. The main difference, therefore, between 1918–21 and 1930–33 consisted in Stalin refusing to back down even after his policies had obviously wreaked havoc in the countryside, had further depressed the urban standard of living, and had not even made possible consistently big exports of grain.

The Bolshevik assault on the peasantry under Stalin ran up against a wall of solidarity in the village. Left to their own devices, most peasants would have presented a united front. Even many workers felt strong bonds with the peasantry, from

◂ Detail of "Are you ready for the spring planting campaign?," *see page 119*.

ДОЛОЙ

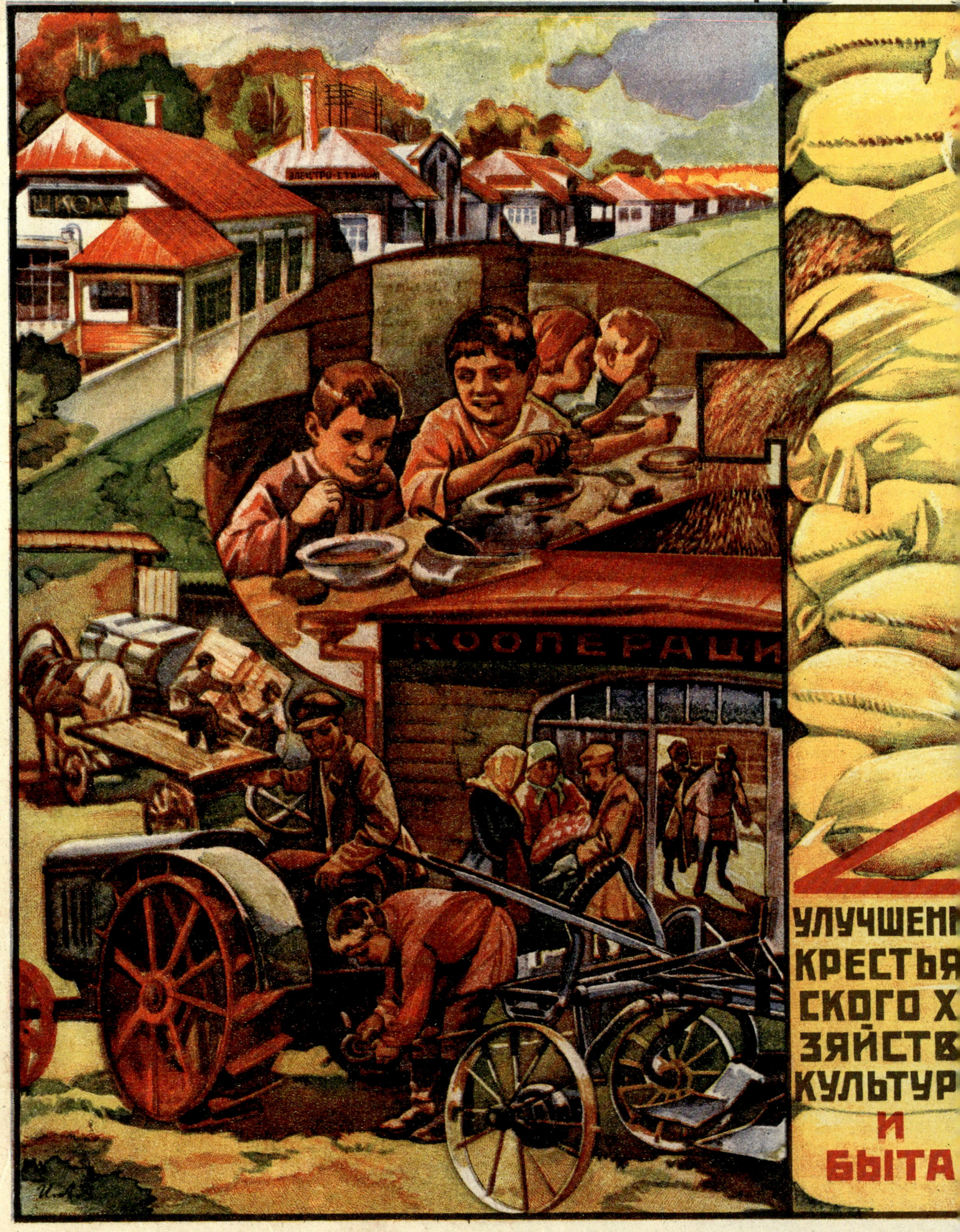

До 80 миллионов пудов хлеба в год перегоняют крестьяне нашей страны на самогон.

Крестьянин! Знаешь ли ты, что самогон отравляет твое здоровье, что ты разоряешь свое хозяйство, что твоя семья голодает: ты отнимаешь у своих детей хлеб на самогон, ты не посы-

лаешь их в школы, тебя з
кулаку, ты помогаешь кула
ты ему даешь хлеб на са

Крестьянин! 80 милли
няешь на самогон, помогли

Главлит № А-27110. Гиз № 29483. Заказ № 3532. Тираж 50 000.

ГОСУДАРСТЕ

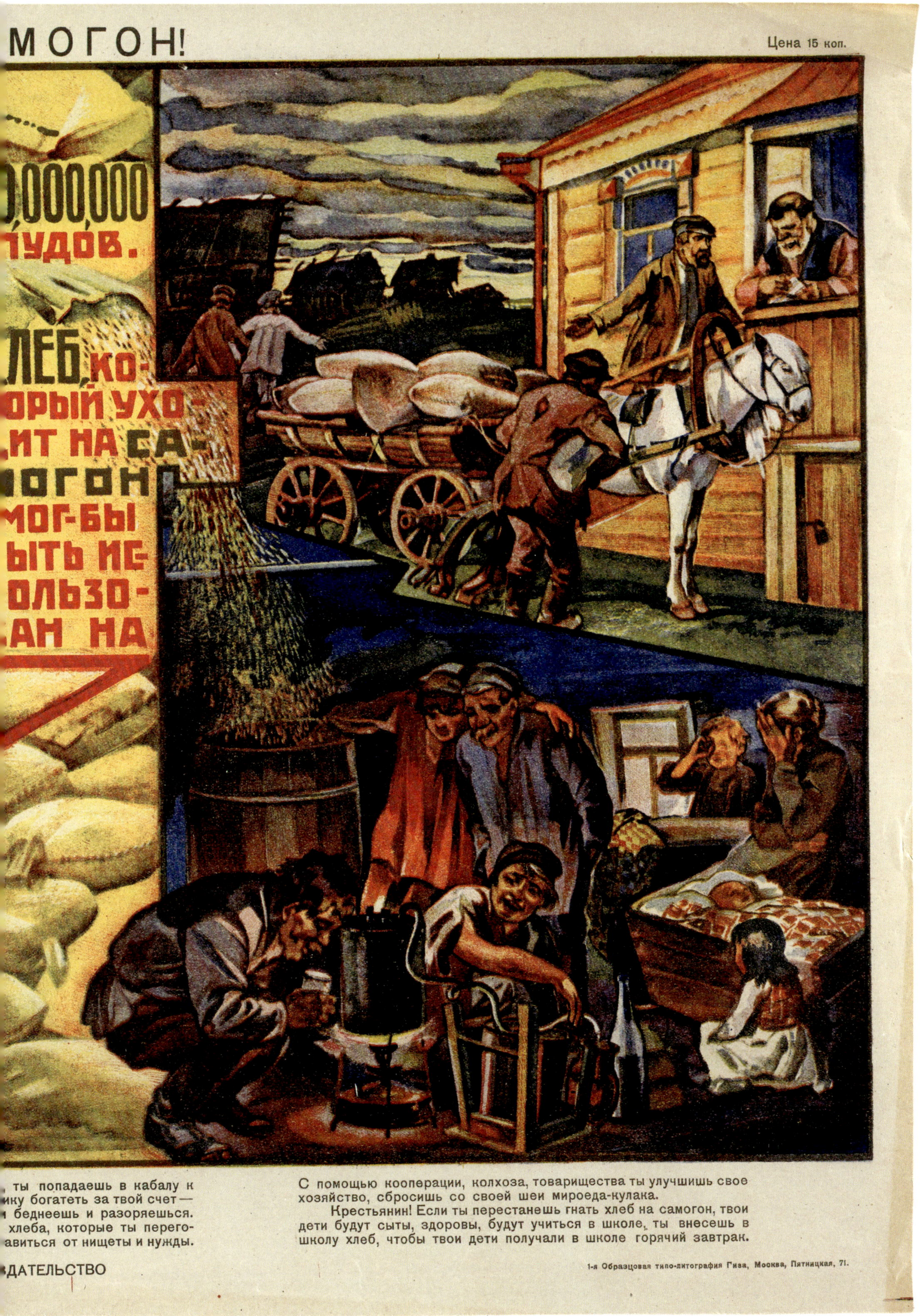

Fig. 60 I.L. "Down with moonshine! Grain used to make home brew could be used for improving peasant agriculture, culture, and way of life." 1930–40?

whose midst most of them had recently emerged. For one thing, their standard of living had fallen precipitously. One Moscow textile factory worker expressed frustration in 1933: "You say that the state and collective farms are sowing more land and increasing the herds of cattle, but provisioning for us is worse and worse. We have stopped believing you. In the first five-year plan, they fed us meat bones, in the second, they will feed us dog meat." Nor did they all buy the argument of "kulak" sabotage. According to a steelworker, "There never were any kulaks in the countryside. They took every last thing from poor people in bast shoes [*lapti*]."[2] In other words, collectivization turned practically the entire country against the Bolsheviks, just as civil war economic policies had done in 1918–20.

So if bigger harvests and larger exports were not achieved, why did the leadership consider the policy a "triumph"? Eminent Western historians have interpreted the motivations behind collectivization in a variety of ways. For Jörg Baberowski, a prolific German scholar who has studied the topic closely in recent years, the policy melded together "the pursuit of an unambiguous, 'European' orderliness with the mad conviction that enemy collectives not fitting into this grandiose vision would have to be destroyed."[3] The vision was "European" in the sense that since the Enlightenment Europeans had sought to bring society, politics, the economy, and human behavior into conformity with rational designs. Lynne Viola concurs, arguing that "special settlement planning in 1930 was socialist realism in another context. It represented an 'imagined future'—laid out in endless plans, reports, memos."[4] And one might add—posters. Or as Kosior expressed the idea on January 12, 1933, the party "proceeded on the assumption that without collectivization, it would be impossible to lead our country onto the highway of constructing the economic foundations of socialism and free millions and millions of working peasants from destitution and ignorance."[5] But did they really mean it?

The chief distinction in Soviet Russia, of course, was the willingness of the leadership to use massive violence and ultimately state terror to force conformity, and not necessarily to rational designs. Nor was Stalin the first Bolshevik to declare "war" on the countryside. During the first Bolshevik assault on the peasantry, Lenin had seen enemies everywhere and had staged show trials to terrorize those enemies, to prove to the population that the Bolsheviks were militant revolutionaries, and in part to deflect attention from the manifest failures of economic policy. The two biggest show trials were furthermore mounted against the two organizations most closely aligned with the peasantry, namely, the Socialist Revolutionary Party and the Russian Orthodox Church. Stalin, of course, carried out his hunt for enemies on an immensely broader scale. The policies of collectivization and "dekulakization" mobilized the entire state apparatus, with the secret police and the party at the forefront, for the purpose of subduing and subordinating a majority of the Soviet population. Massive violence, rejection of compromise, guilt by association, pitting of population groups, individuals, and families against each other—in a word: total war. Stalin's primary target even in 1937, at the height of the Great Terror, was still first and foremost the peasantry, or at least its most menacing fraction, the "kulaks."

Yet were Stalin and his associates taking control of the countryside to shape it into an aesthetically beautiful, rationally modern landscape? The facts that the government devoted few resources and little attention to the details of collective farm life and that few senior officials actually ventured

Fig. 61 A. Kokorekin. "Are you ready for the spring planting campaign?" 1932.

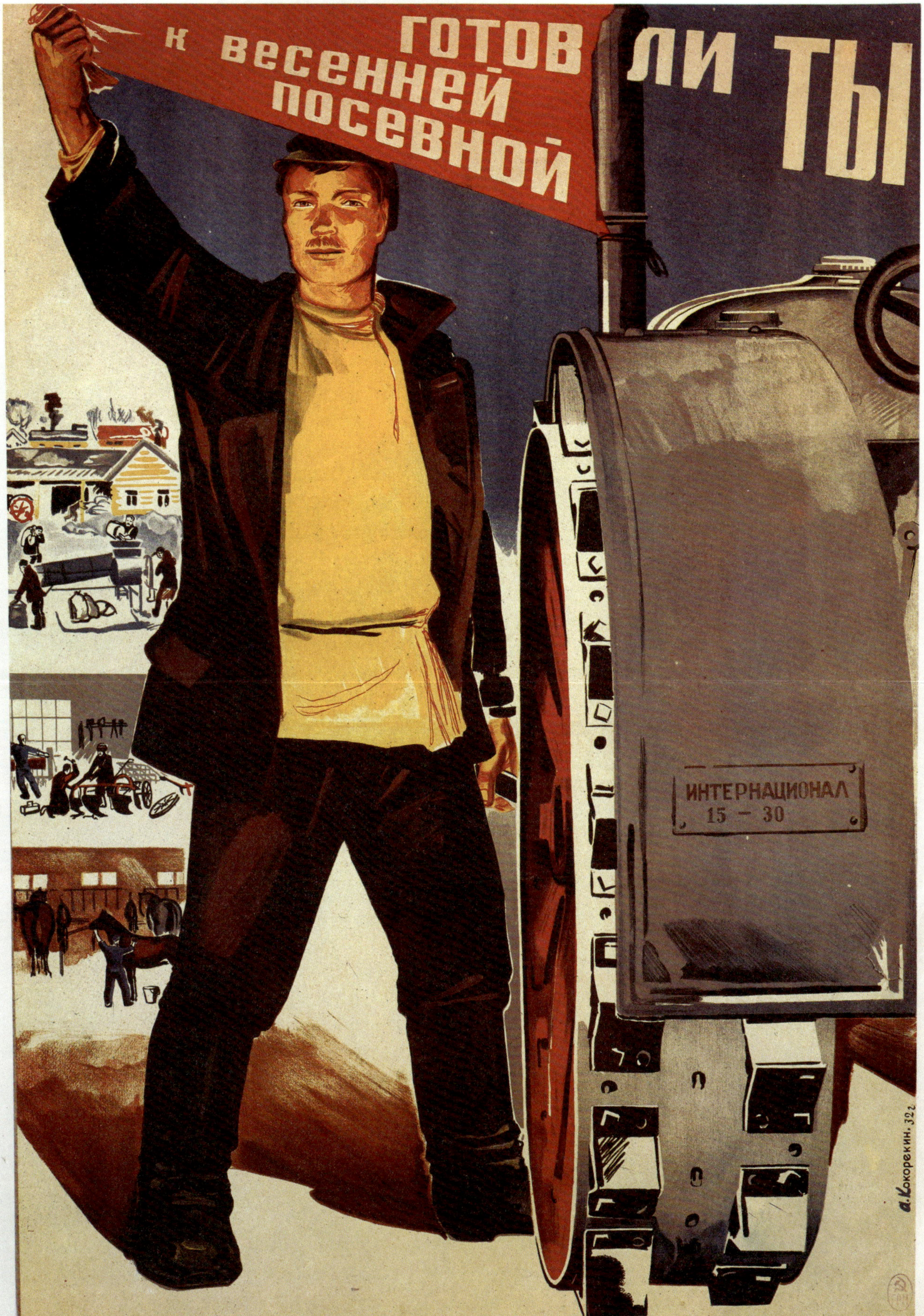
Готов ли ты
к весенней посевной
ИНТЕРНАЦИОНАЛ
15 — 30
А. Кокорекин. 32 г.
Цена 30 коп.
ОГИЗ—ИЗОГИЗ
Москва—Ленинград

into the villages after the fundamentals of collectivization had been achieved—Stalin personally seems not to have returned to the countryside after his foray in early 1929—suggest otherwise. The best clue to their intentions was Kaganovich's exclamation in 1935 about "our absolutely unprecedented, stupefying victory, a victory of Stalinism." That is to say, getting hold of the grain, whatever the cost, whatever the damage to the agricultural system, was the paramount goal. In a speech on January 7, 1933, Stalin listed five principal agricultural achievements of the first Five-Year Plan: the formation of over two hundred thousand collective farms, the expansion of crop area by 21 million hectares, the drawing of over 60 percent of peasant households into collective farms, the routing of the "kulaks," and—the only achievement that gave the regime anything positive—the "procurement of 1,200 to 1,400 million poods of marketable grain annually, instead of the 500,000,000–600,000,000 poods that were procured in the period when individual peasant farming predominated."[6] In other words, the state had succeeded in seizing over 25 million tons of grain out of a harvest of some 70 million tons, according to official estimates in January 1933, or over 35 percent of the total output, in exchange for the most meager compensation. Collectivization achieved nothing more than that. Given the fact that the average harvests both before World War I and in the years 1927–28 were somewhat higher than the average for 1933–38, despite a higher population (just over 160 million in 1910 and just over 170 million in 1939), the Soviet people, especially in the countryside, could not have been eating their fill. And much of what they ate they grew on their family garden plots.

The great economic dilemma of the 1920s, around which Lenin's potential successors danced in mortal combat, had been how to get sufficient grain from the countryside to feed the growing urban population and subsidize industrialization without paying enough to make peasants want to sell. The goal, ultimately, was to defy the law of supply and demand. According to Andrzej Walicki, such precisely was the essence of Marxism from its founding: the repudiation not so much of capital and private property, but of the market.[7] What is the market? It is a mechanism for regulating economic relations without a central arbiter, a way for individuals and groups to come to economic terms on the basis of the best interest of each party, as interpreted by those parties—trillions of times each day. To Marx and his followers, such a mechanism appeared messy, irrational, chaotic. Economic relations under socialism would be determined not by some "invisible hand" secretly guiding the transactions of millions but by rational organization and, ultimately, planning. Of course, millions of peasant producers can typically decide better what to plant, when to harvest, how much—and when and where and for how much—to sell, as well as kindred economic questions. Certainly they can judge such matters better than political operatives with little or no agronomic experience or knowledge. But for intellectuals fanatically committed to abstract ideals, such practicalities counted for little. It helped the appeal of Marxism that its dogmas connected with the deep-seated human resentment of people who are better-off. Stalin's entrenched suspicion and even paranoia, as well as the tense international situation, also contributed. Taken together, such goals, ideals, pathologies, and legitimate worries kept Stalin pushing forward, even after horrific disaster began to strike in 1932, producing the worst famine in modern Russian history.

In fact, the Great Famine of 1932–33 was one of the worst catastrophes of the catastrophe-laden twentieth century. With as many as 7 million "excess deaths," to use the jarring demographic terminology, the episode resulted in a higher level of mortality in a short time span in a single country than any other human cataclysm of the age, except China's Great Famine of 1958–62, which resulted, according to a scholar with access to archival documentation, in some 45 million deaths.[8] If one reckons the Chinese population to have been roughly 4 times greater than that of the USSR (roughly 600 million versus 160 million) and the time span to have been at least twice as long, however, then the proportion of excess yearly fatalities to population size was actually fairly similar between the two tragedies. Certainly the fact that both instances resulted from forced collectivization carried out by Communist governments according to Marxist principles suggests a clear affinity. Both famines, in other words, were largely man-made calamities.

Somewhat less clear, however, is the question of intentionality. Did Stalin and the Communist Party leadership consciously use the famine to punish—or even destroy—the Ukrainian people specifically? Did they, in other words, engage in genocide? In 2006, the Ukrainian parliament, the Verkhovna Rada, passed a law denouncing the famine in Ukraine, the Holodomor (or politically motivated famine), as genocide. Canada, along with a few other countries, passed analogous resolutions.[9] Later that year, the Russian legislature, the Duma, adopted a declaration denying that the tragedy of 1932–33 had any Ukrainian specificity. The Russian government did, however, support a 2010 resolution of the Parliamentary Assembly of the Council of Europe proclaiming that "millions of innocent people in Belarus, Kazakhstan, Moldova, Russia and Ukraine, which were parts of the Soviet Union, lost their lives as a result of mass starvation caused by the cruel and deliberate actions and policies of the Soviet regime."[10] Such an assessment paralleled that of the best Russian experts on the topic, like Viktor Kondrashin. In an interview in 2008 with the newspaper *Izvestiia*, he reasoned that "Russia wasn't killing Ukraine. The leader [*vozhd'*] was killing his people. Debates about who suffered more from the Stalinist regime," he warned, "are immoral and politically dangerous."[11]

There are indeed a number of factors to consider. As noted above, a larger proportion of Kazakhs died in the Great Famine than Ukrainians. Since 1918, and on a theoretical basis going back to Marx's disparagement of the peasantry, the Bolsheviks had been deeply hostile to agricultural workers, seeing them as backward, superstitious, ignorant, and inherently individualistic and acquisitive. Even the alliance (*smychka*) with the peasantry that Lenin advocated was always meant to be temporary and implied a raising of the rural consciousness to the level of proletarians. Most peasants remained fairly neutral during the civil war, leaning on average slightly toward the Bolsheviks. But major agrarian rebellions, fiercely crushed in 1921 and 1922, served the leadership notice that the peasantry was its most menacing potential opponent. It also happened that many of the biggest grain-producing regions, including Ukraine and the North Caucasus, were home to large numbers of ethnic minorities. Additionally, religious faith—another direct threat to the totalizing Marxist faith—was probably strongest in the countryside. Altogether, then, one can see that even on purely rational and pragmatic grounds, the Bolsheviks had the most to fear from rural dwellers. Ukraine, moreover, had the biggest agricultural population of the country—according to the 1926 census, there were 24 million rural inhabitants in Ukraine but only 22 million in the Russian Republic.[12] So, when collectivization began in early 1930 and

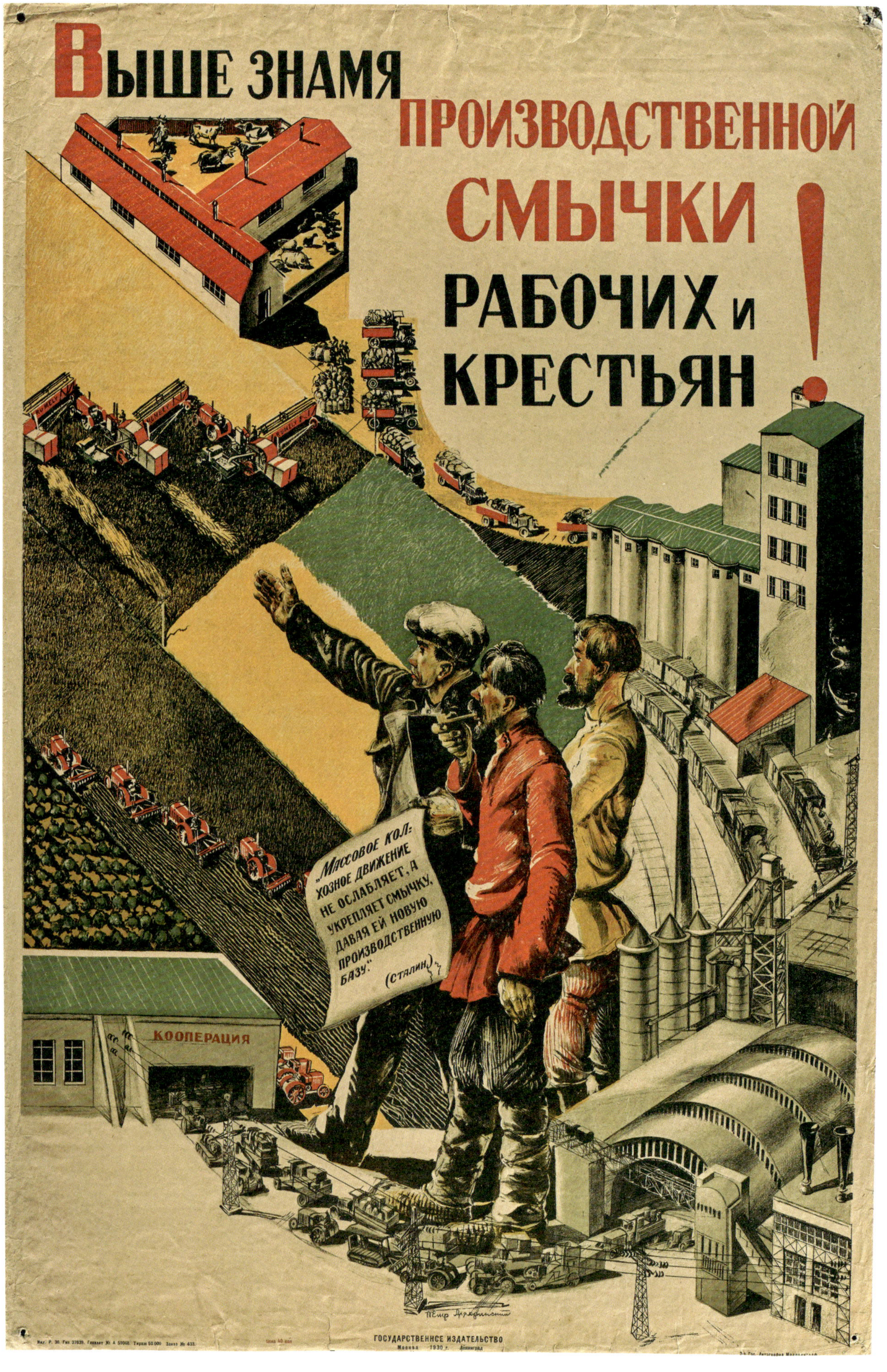
Выше знамя
производственной
смычки
рабочих и
крестьян
!
„Массовое кол-
хозное движение
не ослабляет, а
укрепляет смычку,
давая ей новую
производственную
базу."
(Сталин)
КООПЕРАЦИЯ
ГОСУДАРСТВЕННОЕ ИЗДАТЕЛЬСТВО
Москва 1930 г. Ленинград

massive unrest broke out across the grain-producing regions, the Bolshevik leaders were acutely concerned. Such concern explains their dramatic temporary retreat, of which Stalin's famous "Dizzy with Success" speech on March 2 was the most obvious sign. The fact that vastly more disturbances with vastly more participants took place in Ukraine in the course of 1930—at least 50 percent of the total for the country—must have worried Stalin and his associates to no end.[13] Teaching the entire Soviet peasantry a lesson, crushing its will, showing to whom the grain belonged, and forcing agricultural toilers to work conscientiously to survive were undoubtedly goals that the famine enabled the regime to accomplish. The leadership may well have used the famine specifically to teach such lessons to the Ukrainian peasantry, but certainly not only to it. Kondrashin has calculated that the highest number of deaths as a proportion of population in 1926–37 occurred in Saratov Oblast, along the Volga River. Many other regions also lost a higher percentage of population than did Ukraine.[14]

Whatever the intentions of the leadership in regard to the famine, total secrecy was maintained. Even in private, only the barest mention of death by starvation was made. Officially it never occurred. Only on September 9, 1940, at a closed special meeting of Communist Party officials in charge of ideology, propaganda, and culture, along with the Presidium of the Union of Writers, convened in the Kremlin to discuss the film *Zakon Zhizni* (*The Law of Life*), which had just come out and was quickly pulled from cinemas, did Stalin admit in passing that "25–30 million people in the past were starving; there was not enough grain."[15] Clearly he was referring to the 1932–33 famine. Such references were extremely scarce. Even in his Secret Speech to the Central Committee denouncing Stalinism in 1956, Khrushchev did not mention the famine or even the widespread starvation that followed universal collectivization.

Tragically, the collective farm system was less efficient and productive than individual peasant farmers using relatively primitive implements before collectivization. In 1933, according to official government statistics, the average grain yield per hectare of land was 8.78 centners on the collective farms, 8.44 on the state farms, and 9.41 on individually held farms.[16] Such a comparison did not even take into account one of the most devastating results of Stalin's agricultural policies: the massive destruction of livestock. At least half of the country's stock of horses, cattle, swine, and sheep was lost—through neglect, wanton slaughter, or unplanned consumption—during the years of forced collectivization and famine. The greater efficiency of individual farming, combined with vastly more animal power and available livestock, would have added up to dramatic gains. The economist Holland Hunter and the econometrician Janusz Szyrmer calculated hypothetically that, if collectivization had not occurred, the total capital stock of the USSR by 1940 would have been 29–37 percent greater and—even more striking—the heavy industry and construction sectors would have grown nearly twice as big.[17] In other words, without collectivization precisely those sectors of greatest concern to the party leadership could potentially have expanded most dramatically and made possible the more effective "building of socialism."

As the above statistics suggest, productivity was worst in what was initially touted with great fanfare as "grain factories"—the state farms. According to the Soviet economist A. A. Barsov, "In 1938, only 12.4 percent of the state farms were profitable since the cost of a centner of grain they produced was 2.5 times higher than what the state paid for it."[18] How could this be? The main reason was

Fig. 62 P.A. Aliakrinsky. "Raise the banner of the productive *smychka* of workers and peasants!" 1930.

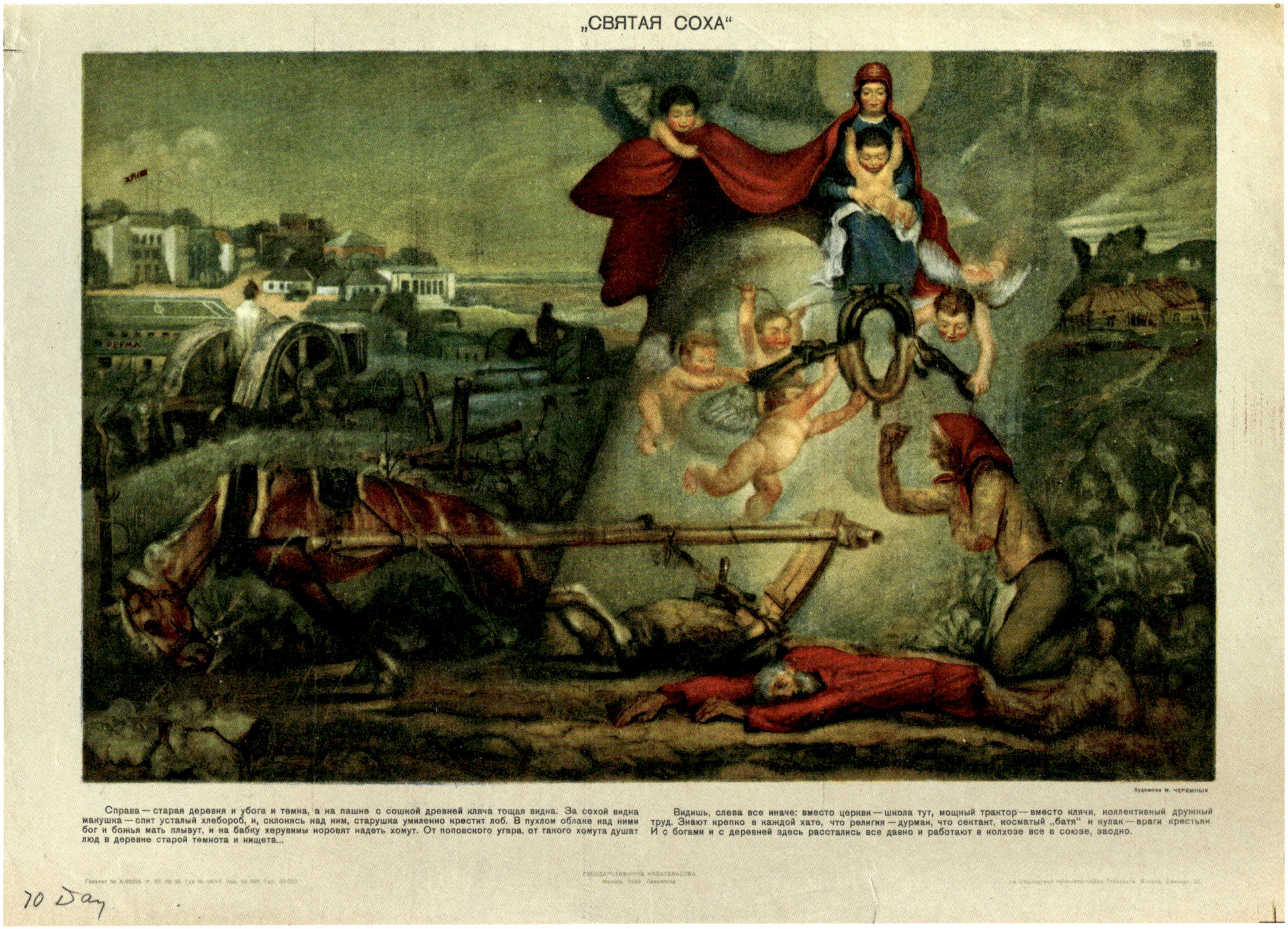

a huge, but largely fruitless, investment in imported farm machinery. Although the state farms cultivated roughly seven times less land than the collective farms, they disposed of only 40 percent fewer tractors (in terms of horsepower).[19] Employees on both state and collective farms on the whole either failed to use such machinery to full advantage or, often, to use it properly. Precious mechanical resources broke down and rusted. As a result, output on the farms did not even pay for the imports of agricultural technology delivered to them.

The collective and especially the much smaller state farm systems were woefully inefficient in comparison to the grandiose expectations placed on them by the party leadership, but how about in international perspective? How did the Soviet collective farmers stack up against the most efficient agricultural workers in the world, those in America? At the height of the Great Depression, in 1938, the United States produced two to four times more meat and roughly 1.5 times more agricultural products overall—with an agricultural workforce roughly three times smaller.[20] After World War II, when the United States had finally pulled out of the Depression, the gap would grow much larger. In 1970,

▲ **Fig. 63** M.M. Cheremnykh. "A sacred plow." 1930. The "sacred wooden plow" contrasted with mechanization, a well-organized collective farm, and a brand-new school.

▶ **Fig. 64** I. Mrachkovsky. "The timely repair of harvesting machines and implements will ensure the successful gathering of the socialist harvest." 1932.

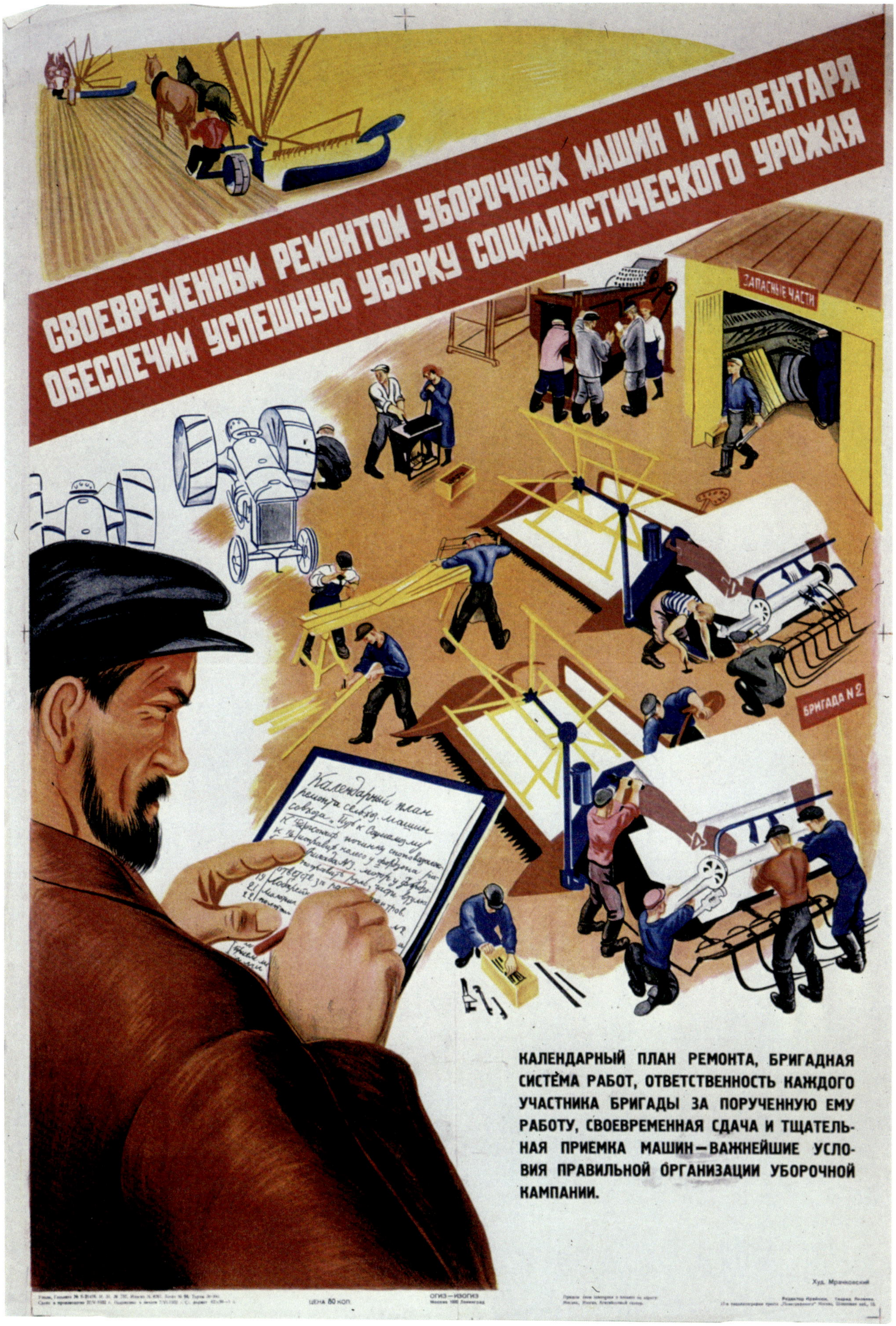
СВОЕВРЕМЕННЫМ РЕМОНТОМ УБОРОЧНЫХ МАШИН И ИНВЕНТАРЯ
ОБЕСПЕЧИМ УСПЕШНУЮ УБОРКУ СОЦИАЛИСТИЧЕСКОГО УРОЖАЯ
ЗАПАСНЫЕ ЧАСТИ
БРИГАДА № 2
КАЛЕНДАРНЫЙ ПЛАН РЕМОНТА, БРИГАДНАЯ СИСТЕМА РАБОТ, ОТВЕТСТВЕННОСТЬ КАЖДОГО УЧАСТНИКА БРИГАДЫ ЗА ПОРУЧЕННУЮ ЕМУ РАБОТУ, СВОЕВРЕМЕННАЯ СДАЧА И ТЩАТЕЛЬНАЯ ПРИЕМКА МАШИН — ВАЖНЕЙШИЕ УСЛОВИЯ ПРАВИЛЬНОЙ ОРГАНИЗАЦИИ УБОРОЧНОЙ КАМПАНИИ.
Худ. Мрачковский
ЦЕНА 80 КОП.
ОГИЗ — ИЗОГИЗ
Москва 1932 Ленинград

for example, some 3,463,000 American farmers produced enough food to feed the entire population and to export vastly more than their 26,439,000 Soviet counterparts (many of whom also worked at least part-time in other industries), whose smaller production often did not prevent shortages from occurring throughout the USSR.[21] Each branch of agriculture was different, but American farmers both before and after the war produced multiple times more food per worker than Soviet collective farmers.

Many policies, trends, practices, and attitudes that first emerged during collectivization became enduring features of Soviet governance and culture. The Soviet security police, which since its founding in December 1917 had played an important role in Soviet political life, henceforth became a central, ubiquitous, and nearly all-powerful player, whether under the name of OGPU, NKVD, or KGB. From the time of collectivization, guilt by association with relatives became an ongoing threat to families and individuals, leading both to many instances of explicit renunciation of family ties and to serious punishment. Of course, the passport system, which emerged for population control during the Great Famine, continued to regulate many aspects of daily life (including residence, access to education, and employment).

The famine itself had a powerful impact on the Soviet population. The tens of millions of people who were starving, who went without food perhaps for days or even weeks, who desperately resorted to eating pets, rats, diseased livestock, the bark from trees, grass, even other human beings—yet lived—probably made as big an impact on Soviet society and culture as the millions who perished. The fact that they suffered such privation and perhaps engaged in shameful acts to survive a famine that most people considered man-made, brought on by government policies, and even possibly intentionally perpetrated against them could not but drastically affect their outlook on life, on the regime, on their society.[22] The deep suspiciousness Soviet citizens felt toward any newcomer, anyone who was not *nash* (ours), surely gained some of its intensity from the close brush of tens of millions with death—and their virtual abandonment by the state—in 1932–33.

A relatively small number of political prisoners were incarcerated in prerevolutionary penitentiaries and jails and, starting in 1923, exiled to the Solovetsky Islands in the White Sea (roughly thirty thousand were counted in 1927). Yet it was collectivization and the anti-"kulak" campaign that gave birth to the gulag—a network of forced-labor camps, compounds, and special settlements eventually containing millions of prisoners under the direct control of the secret police. The number of people sentenced to death, camps, exile, and other punishments for counterrevolutionary crimes in 1930–36, according to a secret memorandum prepared in 1953 after Stalin's death, totaled 1,391,093, yet this figure did not include the many millions executed or deported as "kulaks" in 1930 and 1931.[23] The extreme violence and terror that came to characterize Stalinism became a conscious and pervasive practice only during forced collectivization, despite some significant antecedents, such as the "war against the peasant bourgeoisie" and the Red Terror of 1918.

The Bolsheviks had persecuted the Church and religious believers from the beginning of their rule, but only during collectivization did they undertake a thoroughgoing onslaught against them, closing the vast majority of houses of worship. Henceforth the Soviet Union had a built environment where few faithful could make their way conveniently to a church. That was only one impediment to regular religious

▸ **Fig. 65** V. S. Ivanov and O. K. Burova. "To advanced collective farms—thank you from the front." 1944.

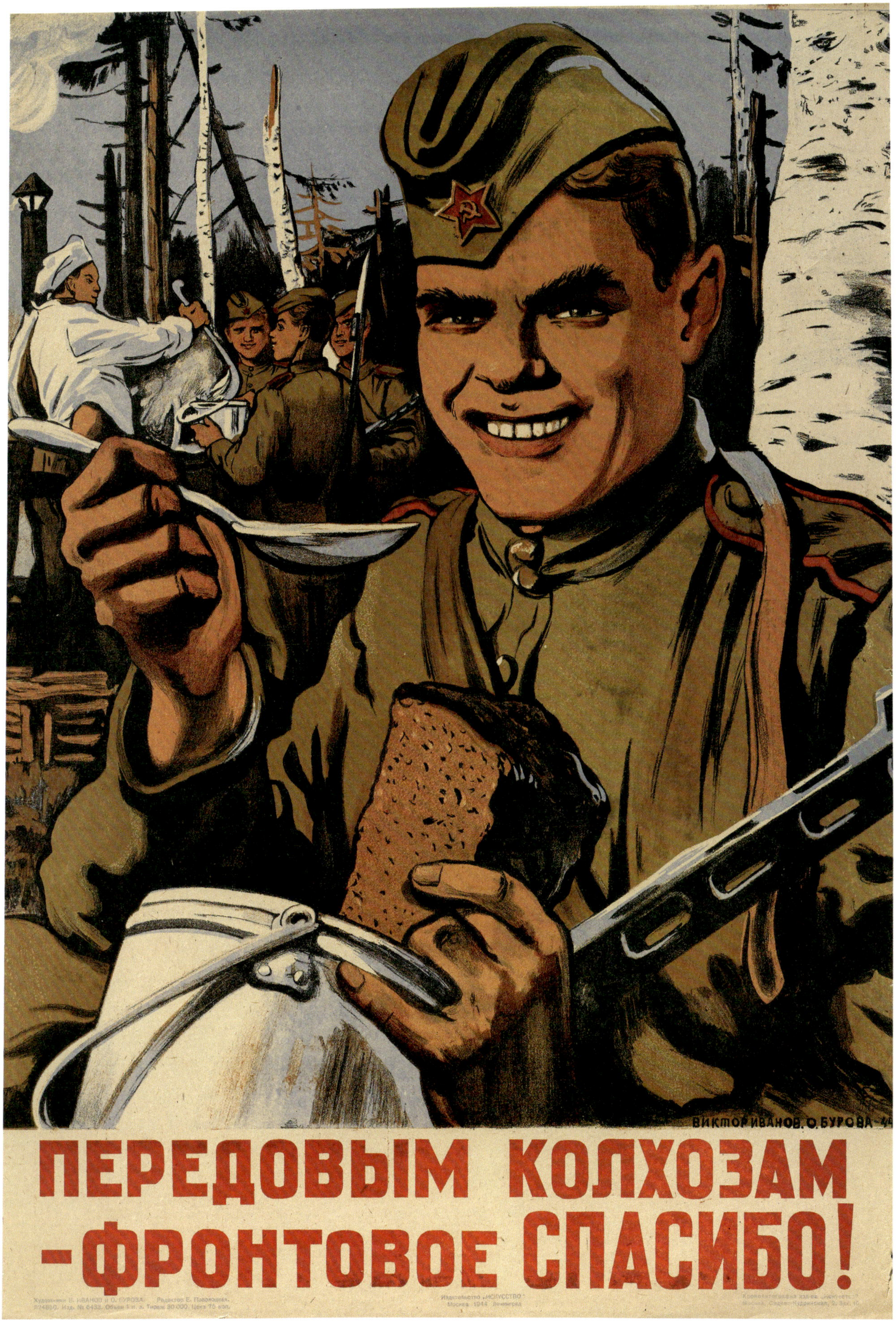
ВИКТОР ИВАНОВ. О. БУРОВА 44
ПЕРЕДОВЫМ КОЛХОЗАМ
-ФРОНТОВОЕ СПАСИБО!
Издательство „ИСКУССТВО"
Москва 1944 Ленинград

practice for most people in the USSR, but it was probably the first major one.

The lifestyle of rural dwellers was established during collectivization for the rest of the Soviet era. Living in the countryside meant doing without a wide variety of amenities, including, for many decades, electricity and running water. In most countries, rural life is generally more "primitive," but the contrast in the USSR was especially stark. Tens of thousands of villages had been destroyed during the fierce combat of World War II. For years afterward, much of the built environment in the countryside remained dilapidated. Nikita Khrushchev came to power with a commitment to both dramatically raising the standard of living in the countryside and urbanizing rural life. The plan worked out in various Moscow offices was to build 120,000 agrocities (*agrogoroda*) throughout all rural regions. The remaining 580,000 villages were to be abandoned.[24] As with so many of his harebrained schemes, almost nothing was to come of the new strategy. Toward the late Soviet era, in fact, most villages were little changed from the postwar era, except that they gradually emptied out of young and dynamic members. Rural alcoholism was particularly rampant in a society plagued with the malaise.

Naturally, the most direct consequence of collectivization was the ruination of the agricultural economy. The former breadbasket of Europe was soon obligated to import large quantities of grain on a nearly annual basis. Many kinds of food and often even bread became scarce commodities, or *defitsitnye produkty* (deficit foodstuffs), in Soviet parlance. The shelves of ordinary food stores were often practically bare, at least of items most people were interested in buying, and things people wanted often required waiting in line.

In other words, it was collectivization more than anything else that set in motion the processes giving the Soviet Union its ultimate shape as a society. Great triumphs there certainly were—in defense, high culture, sports, space, chess, mathematics, and theoretical science—but these were achieved in the most advanced urban settings. The havoc and destruction wrought in the countryside in 1930–33 sowed the seeds of destruction of the entire Soviet experiment and ultimately rendered those achievements hollow.

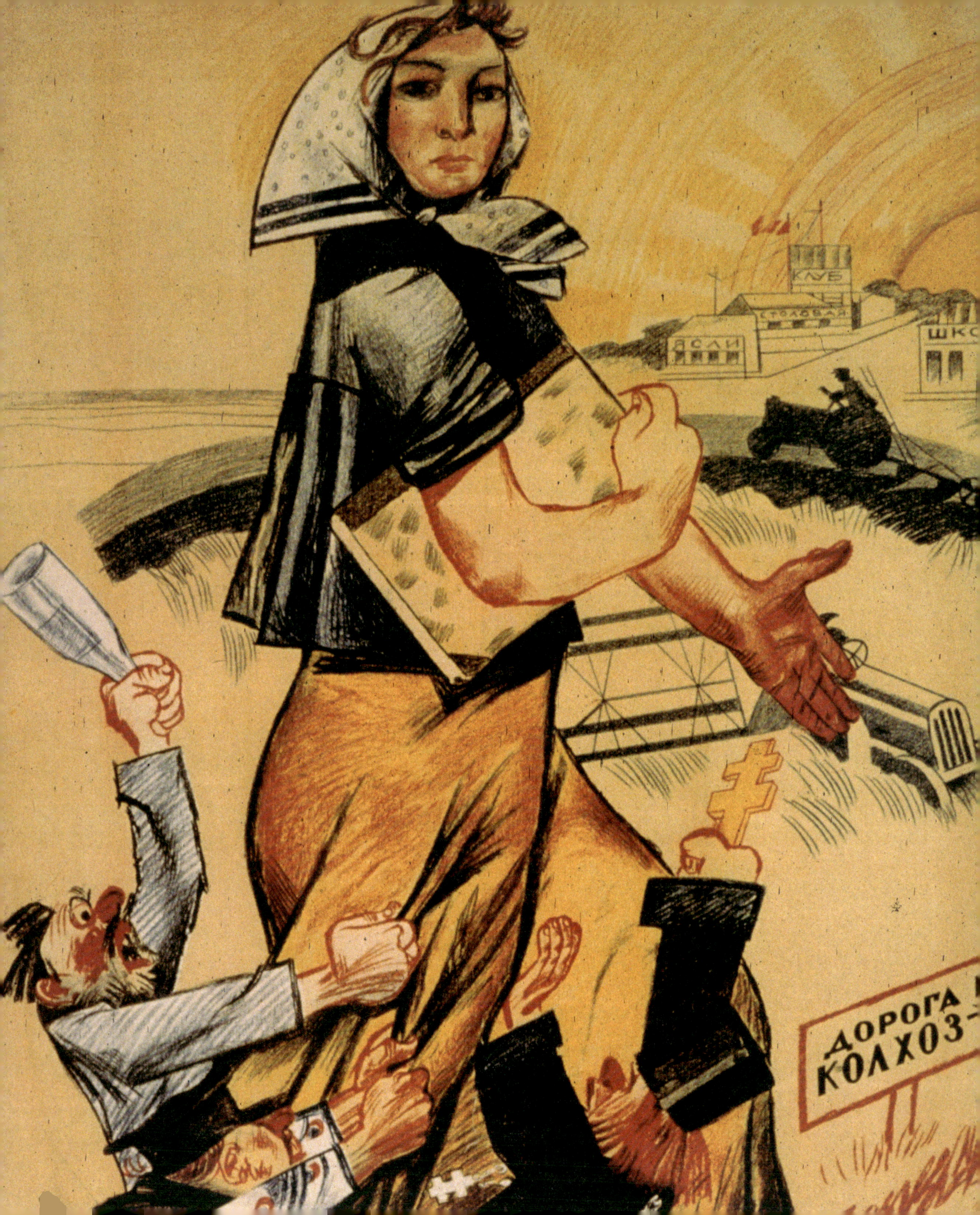
КЛУБ
СТОЛОВАЯ
ЯСЛИ
ДОРОГА
КОЛХОЗ-

CHRONOLOGY OF EVENTS

	1917
February	Tsar abdicates
Summer	Spontaneous land reform
October	Bolsheviks seize power
December	Cheka (secret police) created
	1918
March	Treaty of Brest-Litovsk
May	Czechoslovak revolt; civil war begins; "war" declared on "peasant bourgeoisie"
September	Red Terror
	1919
January–March	Massacre of Cossacks
	1920
	Civil war ends
August	Tambov Uprising
	1921
March	Kronstadt Rebellion; New Economic Policy declared; ban on party factions
	1922
	Famine peaks
February	GPU formed
Summer	Show trials against church and Socialist-Revolutionary leaders
	1923
November	OGPU formed
	1924
January	Lenin dies
August	Peasant rebellion in Georgia
October	Face to the Village campaign proclaimed
December	Stalin proclaims "socialism in one country"

◀ Detail of "Peasant women, join the collective farm!," *see page 47*.

1925

April	Bukharin urges peasants to "enrich yourselves"
December 10	Statistical Administration placed under political control
December 18–31	Fourteenth Party Congress; Stalin advocates rapid industrialization

1926

	First census of the USSR; industrial production slightly exceeds the prewar level
June 18	Special tax on profits of NEPmen
August 19	OGPU given orders to help collect grain
December	Official overestimate of grain production (900 million poods, or 16 million tons)

1927

	Creation of tractor commune in the Odessa region provides model for MTS in 1929
	Poor harvests and low grain prices result in lower grain deliveries
January 11	Creation of state grain reserve
March 11	Glavlit via OGPU forbids any mention of problems with grain supply
April 12	Slaughter of Chinese Communists in Shanghai
April 18–26	Fourth Congress of Soviets approves principle of Five-Year Plan
May 27	UK breaks off diplomatic relations with the USSR
June 7	Soviet plenipotentiary representative in Poland, Petr Voikov, killed; war scare
July 6	Start of anti-counterrevolution campaign in countryside
September 12	Glavlit via OGPU forbids reporting on export of grain
November 12–14	Plenum expels Trotsky, Kamenev, and Zinoviev from party
December 2–19	Fifteenth Party Congress approves gradual collectivization, further limits on "kulaks"
December 24	Politburo decree on grain collection

1928

January 4	OGPU agents ordered to get directly involved with grain collection
January 5	First harsh Stalinist decree on grain collection
January 6	Party apparatus ordered to mobilize for grain collection
January 14	Party directive calls "kulak" enemy of Soviet power; Stalin leaves for Siberia
January 17	Trotsky exiled to Central Asia
January 18–February 4	Stalin travels to Siberia: grain requisitioning, arrest of resistors, attacks on "kulaks"
February 4	OGPU cable to local organs: Telegraph results of mass operations
April 21	Taxes increased on peasants
May 18–July 6	Shakhty trial
June 4–12	Central Committee Plenum; rightists criticize Stalin's economic policy
July	Plenum lessens repression, rehabilitates many peasant exiles
September 14	"Milling levy" introduced
September 30	Bukharin again affirms need for union with peasantry in *Pravda*
October 1	Official commencement of first Five-Year Plan; 78 percent of capital investment in industry
December 15	New agricultural codex radically limits rights of peasants

1929

	Magnitogorsk construction begins; return to rationing of bread in cities
January 11	Yearly plan raised by 43 million poods
February	Trotsky expelled from USSR
April 8	Order on tighter control over parishes
April 16–23	Central Committee Plenum condemns right deviation, weakens Bukharin
April 23–29	Five-Year Plan targets increased; fivefold tax penalty and Ural-Siberian method officially sanctioned
June 5	Order to create MTS network
July 18	Order to remove "kulaks" from collective farms
June 27	Fivefold fine directive and law
August 26	Sunday abolished as day of rest
September 9	Peasants were forbidden to sell grain on free market
October 3	Diplomatic relations restored with UK
October 3	OGPU/NKIu order to use all measures up to death for "kulak" terror
November 7	Stalin's "A Year of Great Change" in *Pravda*
November 10–17	Plenum speeds up pace of collectivization; Bukharin expelled from Politburo
December 5	Central Committee commission created to plan collectivization
December 5–10	First congress of shock work brigades; Bukharin removed from Politburo
December 21	Fiftieth birthday of Stalin; start of cult of personality
December 27	Stalin announces universal collectivization and "liquidation of kulaks as class"

1930

January 5	Collectivization to be voluntary and in three stages
January 30	"Dekulakization" degree
March 2	Stalin's "Dizzy with Success" in *Pravda*
March 14	Order of Central Committee on distortions in *kolkhoz* movement
June 26–July 3	Sixteenth Party Congress confirms rapid industrialization
July 30	Peasant land society abolished in districts of universal collectivization

1931

March 6	Congress of Soviets introduces "labor day" for *kolkhozniki*
July 20	Rooting out of "kulaks" proclaimed completed
October 11	Decision on complete liquidation of private trade

1932

	Collective farm markets open
April 23	Doctrine of socialist realism adopted as official
August 7	Law on socialist property
November 8	Suicide of Nadezhda Allilueva, Stalin's second wife
November 15	Tightened labor discipline: layoff for one absence
December 27	Internal passports instituted
December 31	Five-Year Plan proclaimed completed in four years

1933

	Famine raging
January 7–12	Party plenum decides to launch party purge, set up political sections in MTS
January 19	Return to system of obligatory deliveries of grain to state by collective farms
January 22	Peasants forbidden to migrate from grain-producing regions
December 23	Collective farmers allowed individual garden plots

1934

May 27	Partial amnesty of former "kulaks"
June 8	Law on treason (death penalty, collective responsibility of family)
July 10	Reorganization of political police; NKVD headed by Yagoda
December 1	Murder of Sergei Kirov

1935

January 1	Rationing ends on flour and bread, but prices increase sharply
February 17	New rules on collective farms (individual plots, livestock)
August 30–31	Stakhanov feat
September 29	Consumer cooperatives in cities abolished
October 1	Rationing of meat, oil, fish, sugar, potatoes ended; prices sharply increased
November 17	Stalin: "life has become better"

1936

Stalin Constitution

Yezhov replaces Yagoda

1937

July	Start of anti-"kulak" operation and Great Terror

1938

October 22	Exiled sixteen-year-old "kulaks" given right to obtain passports

1939

May	Plenum: "defense of collective farm property from dissipation"

NOTES

PREFACE

1 The scholarly titles relating to "anti-peasant repression" published in Russia alone filled a 200-page book in 2010. See M. G. Stepanov, *Otechestvennaia istoriografiia antikrest'ianskikh repressii v SSSR, 1929–1933 gg.* (Abakan: Izd. Khakasskogo gos. univ. im. N. F. Katanova, 2010).

INTRODUCTION

1 Edmond Théry, *La transformation économique de la Russie* (Paris: Économiste Européen, 1914), xvii.

2 As Kalinin's remark suggested, the term *kulak* did not describe a meaningful social distinction. It will therefore be used only in quotation marks (by the author except for the cited works).

3 Most government officials listed as passing away in 1936–39 perished in the Great Terror, when, according to official numbers, 681,692 people were executed.

4 As quoted in Stephen Kotkin, *Stalin*, vol. 1, *Paradoxes of Power, 1878–1928* (New York: Penguin Press, 2014), 570.

5 As quoted in Alexandre Livchine, "Les campagnes russes et l'autorité locale: 'Lettres de l'interieur' à l'époque de la NEP," *Communisme*, nos. 42/43/44 (1995): 104.

6 As quoted in Aleksandr Livshin and Igor' Orlov, eds., *Vlast' i obshchestvo: Dialog v pis'makh* (Moscow: ROSSPEN, 2002), 65.

7 Statistics in R. W. Davies, *The Industrialization of Soviet Russia*, vol. 1, *The Socialist Offensive: The Collectivization of Soviet Agriculture, 1929–1930* (Cambridge, MA: Harvard University Press, 1980), 25.

8 Soviet statistics in V. P. Danilov, *Sovetskaia dokolkhoznaia derevnia, sotsial'naia struktura, sotsial'nye otnosheniia* (Moscow: Nauka, 1979), 305. American statistic in *Agriculture Fact Book 1997* (Washington, DC: US Department of Agriculture, 1997), 24.

9 J. V. Stalin, Political Report of the Central Committee to the Fifteenth Congress of the C.P.S.U.(B.), December 3, 1927, in *Works* (Moscow: Foreign Languages Publishing House, 1952–55), 10:319.

CHAPTER 1

1 His boss, Viacheslav Menzhinsky (1874–1934), being chronically ill, usually left Iagoda in charge.

2 As quoted in Roberta T. Manning, "The Rise and Fall of the Extraordinary Measures, January–June, 1928: Towards a Reassessment of the Onset of the Stalin Revolution," *Carl Beck Papers in Russian and East European Studies* 1504 (January 2001): 9.

3 As quoted in V. A. Il'inykh, *Khronika khlebnogo fronta: Zagotovitel'nye kampanii kontsa 1920-kh v Sibiri* (Moscow: ROSSPEN, 2010), 129.

4 As quoted from V. P. Danilov, "Vvedenie (Istoki i nachalo derevenskoi tragedii)," in *Tragediia sovetskoi derevni: Kollektivizatsiia i raskulachivanie: Dokumenty i materialy v 5 tomakh, 1927–1939*, ed. V. P. Danilov et al. (Moscow: ROSSPEN, 1999), 1:36–37.

5 Statistic supplied in Danilov, "Vvedenie," 1:39.

6 As quoted in Viktor Danilov and Alexis Berelowitch, eds., "Les documents des VChK-OGPU-NKVD sur la campagne soviétique, 1918–1937," *Cahiers du monde russe* 35, no. 3 (July–September 1994): 654.

7 Statistic quoted in Manning, "The Rise and Fall of the Extraordinary Measures," 12.

8 Statistics from Moshe Lewin, *Russian Peasants and Soviet Power: A Study of*

Collectivization, trans. Irene Nove with John Biggart (Evanston, IL: Northwestern University Press, 1968), 228.

9 Nikolai Pogodin, "Sud v derevne Drokinoi," *Pravda* (March 16, 1928), quoted in V. P. Danilov and N. A. Ivnitskii, eds., *Dokumenty svidetel'stvuiut: Iz istorii derevni nakanune i v khode kollektivizatsii, 1927–1932 gg.* (Moscow: Izd. politicheskoi literatury, 1989), 115.

10 Stalin, *Works*, 11:67.

11 "Soviet War on Peasantry. 'Destroying' the Well-to-Do," *Times* (London), April 23, 1928, 14.

12 M. Levshin, "Kulak lovchitsia," *Bednota* (April 28, 1928), quoted in Danilov and Ivnitskii, eds., *Dokumenty svidetel'stvuiut,* 154.

13 Statistics in N. A. Ivnitskii, *Repressivnaia politika sovetskoi vlasti v derevne, 1928–1933* (Moscow: Institut Rossiiskoi Istorii RAN, 2000), 48.

14 Statistics quoted in Hugh D. Hudson, Jr., *Peasants, Political Police, and the Early Soviet State: Surveillance and Accommodation under the New Economic Policy* (New York: Palgrave Macmillan, 2012), 99; N. A. Ivnitskii, *Golod 1932–1933 godov v SSSR: Ukraina, Kazakhstan, Povolzh'e, Tsentral'no-Chernozemnaia oblast', Zapadnaia Sibir', Ural* (Moscow: Sobranie, 2009), 25.

15 Lewin, *Russian Peasants and Soviet Power,* 242.

16 Stalin, *Works*, 11:167.

17 See Hudson, *Peasants, Political Police, and the Early Soviet State*, 107.

18 Statistics from Lewin, *Russian Peasants and Soviet Power*, 216.

19 As quoted in Lynne Viola et al., eds., *The War against the Peasantry, 1927–1930: The Tragedy of the Soviet Countryside* (New Haven, CT: Yale University Press, 2005), 68.

20 Statistics from Viola et al., eds., *The War against the Peasantry*, 68.

21 As quoted in Il'nykh, *Khronika khlebnogo fronta*, 206.

22 As quoted by James Hughes, "Capturing the Russian Peasantry: Stalinist Grain Procurement Policy and the 'Ural-Siberian Method,'" *Slavic Review* 53, no. 1 (1994): 86.

23 As quoted in Ivnitskii, *Repressivnaia politika*, 52.

24 Statistic in Hughes, "Capturing the Russian Peasantry," 101.

25 Quoted by N. A. Ivnitskii, *Klassovaia bor'ba v derevne i likvidatsiia kulachestva kak klassa* (Moscow: Nauka, 1972), 106.

26 As quoted in Nicolas Werth, "Le pouvoir soviétique et l'Église orthodoxe de la collectivisation à la Constitution de 1936," *Revue d'études comparatives Est-Ouest* 24 (1993): 42.

27 As quoted in Werth, "Le pouvoir soviétique," 43.

28 Statistic in Deborah Fitzgerald, "Blinded by Technology: American Agriculture in the Soviet Union, 1928–1932," *Agricultural History* 70, no. 3 (Summer 1996): 468.

29 As quoted in Naum Jasny, *The Socialized Agriculture of the USSR: Plans and Performance* (Stanford, CA: Stanford University Press, 1949), 27. Italics in the original.

30 As quoted in R. W. Davies, *The Industrialization of Soviet Russia*, 1:128.

31 Statistics in Ivnitskii, *Repressivnaia politika*, 61, 70.

32 As quoted in Davies, *The Industrialization of Soviet Russia*, 1:94.

33 Statistics in Davies, *The Industrialization of Soviet Russia*, 1:133.

34 As quoted by Roger R. Reese, "Red Army Opposition to Forced Collectivization, 1929–1930: The Army Wavers," *Slavic Review* 55 (Spring 1996): 39.

35 Stalin, *Works*, 12:134, 137, 138. Italics in the original.

36 Quoted from Ivnitskii, *Golod 1932–1933 godov v SSSR*, 18.

37 Stalin, *Works*, 12:147. Italics in the original.

38 As quoted in Wendy Goldman, "Hunger and Politics: The Prelude to the Great Terror," in *Hunger and Scarcity under State-Socialism*, ed. M. Middell and F. Wemheuer (Leipzig: Leipziger Universitätsverlag, 2012), 33.

CHAPTER 2

1 See Ivnitskii, *Golod 1932–1933 godov v SSSR*, 71–72.

2 As quoted in N. A. Ivnitskii and V. G. Makurov, eds., *Iz istorii raskulachivaniia v Karelii, 1930–1931 gg.: Dokumenty i materialy* (Petrozavodsk: Kareliia, 1991), 17.

3 Statistic in Ivnitskii, *Repressivnaia politika*, 137.

4 As quoted in Ivnitskii and Makurov, eds., *Iz istorii raskulachivaniia v Karelii*, 18.

5 "Soviet and Peasants," *Times* (London),

January 23, 1930, 12; "Persecution of Peasants," *Times* (London), January 27, 1930, 12.

6 Statistics in A. A. Rakov, *Sotsial'no-ekonomicheskie aspecty "raskulachivaniia" krest'ian iuzhnogo Urala, 1930–1934 gg.* (Moscow: Maks-Press, 2012), 119–20.

7 Statistics in Werth, "Le pouvoir soviétique," 44.

8 Statistic in Richard L. Hernandez, "Sacred Sound and Sacred Substance: Church Bells and the Auditory Culture of the Russian Village during the Bolshevik 'Velikii Perelom,'" *American Historical Review* 109 (2004): 1500.

9 See Werth, "Le pouvoir soviétique," 41.

10 Statistic in Ivnitskii, *Repressivnaia politika*, 159.

11 See Manning, "The Rise and Fall of the Extraordinary Measures," 1, 33.

12 Quoted in L. E. Zelenin, "'Revoliutsiia sverkhu': Zavershenie i tragicheskie posledstviia," *Voprosy istorii* 10 (1994): 28–29.

13 Stalin, *Works*, 12:204.

14 Statistics from Lynne Viola, *The Unknown Gulag: The Lost World of Stalin's Special Settlements* (Oxford: Oxford University Press, 2007), 62.

15 Viola, *The Unknown Gulag*, 74, 78.

16 As quoted in Lynne Viola, "Stalin's Empire: The Gulag and Police Colonization in the Soviet Union in the 1930s," in *Stalin and Europe: Imitation and Domination, 1928–1953*, ed. Timothy Snyder and Ray Brandon (New York: Oxford University Press, 2014), 29. Italics in original.

17 Statistics in Rakov, *Sotsial'no-ekonomicheskie aspekty "raskulachivaniia" krest'ian*, 84.

18 Petro G. Grigorenko, *Memoirs*, trans. Thomas P. Whitney (New York: W. W. Norton, 1982), 36.

19 Statistic from Viktor Kondrashin, *Golod 1932–1933 godov: Tragediia rossiiskoi derevni* (Moscow: ROSSPEN, 2008), 71.

20 As quoted in Rakov, *Sotsial'no-ekonomicheskie aspecty "raskulachivaniia" krest'ian*, 86. Ellipses and italics in original.

21 As quoted in Danilov and Berelowitch, eds., "Les documents des VChK-OGPU-NKVD," 669.

22 Stalin, *Works*, 12:343.

23 Statistic supplied in Ivnitskii and Makurov, eds., *Iz istorii raskulachivaniia v Karelii*, 102.

24 Statistics in Ivnitskii, *Golod 1932–1933 godov v SSSR*, 91.

25 Statistic available in R. W. Davies, *The Industrialization of Soviet Russia*, vol. 2, *The Soviet Collective Farm, 1929–1930* (Cambridge, MA: Harvard University Press, 1980), 58.

26 Statistic in Davies, *The Industrialization of Soviet Russia*, 1:380.

27 Quotations from Stephan Merl, *Bauern unter Stalin: Die Formierung des sowjetischen Kolchossystems, 1930–1941* (Berlin: Duncker & Humblot, 1990), 94.

28 As quoted in Ivnitskii and Makurov, eds., *Iz istorii raskulachivaniia v Karelii*, 117.

29 Terms quoted in R. W. Davies and Stephen G. Wheatcroft, *The Years of Hunger: Soviet Agriculture, 1931–1933* (New York: Palgrave Macmillan, 2004), 25.

30 As quoted in Ivnitskii and Makurov, eds., *Iz istorii raskulachivaniia v Karelii*, 120.

31 Statistic taken from N. A. Ivnitskii, *Sud'ba raskulachennykh v SSSR* (Moscow: Sobranie, 2004), 26.

32 Statistic in Davies and Wheatcroft, *The Years of Hunger*, 60.

33 Each task was assigned a specific number of labor days, not necessarily corresponding to the amount of labor required (for example, spreading fertilizer by hand on 0.15 hectare —0.37 acre—earned 1¼ labor days). The more skilled the worker, the more labor days one earned for each task. Thus, tractor drivers earned twice as many labor days for a given task as milkmaids. Payment for each labor day was usually in cash and kind (in the range of one kilogram of grain and 60 kopeks per labor day). It also varied regionally, even farm to farm, and year by year, depending on the surplus of the particular collective farm.

34 Statistics obtained from Ivnitskii, *Sud'ba raskulachennykh*, 250.

35 Statistic in Davies and Wheatcroft, *The Years of Hunger*, 37.

36 Statistics obtained from Ivnitskii, *Sud'ba raskulachennykh*, 271.

37 Statistics in V. N. Zemskov, "'Kulatskaia ssylka' v 30-e gody," *Sotsiologicheskie issledovaniia* 10 (1991): 10.

38 Statistic in Davies and Wheatcroft, *The Years of Hunger*, 4.

39 As quoted in Viola, *The Unknown Gulag*, 113.

40 As quoted in R. W. Davies, "Making Economic Policy," in *Behind the Façade of*

Stalin's Command Economy: Evidence from the Soviet State and Party Archives, ed. Paul R. Gregory (Stanford, CA: Hoover Institution Press, 2001), 70–71.

41 "Real Russia," *Times* (London), October 12, 1931, 13.

42 Statistics in Kondrashin, *Golod 1932–1933 godov*, 88.

43 Statistics in Davies, *The Industrialization of Soviet Russia*, 1:361.

44 Statistic cited in Sheila Fitzpatrick, "The Great Departure: Rural-Urban Migration in the Soviet Union, 1929–1933," in *Social Dimensions of Soviet Industrialization*, ed. William R. Rosenberg and Lewis H. Siegelbaum (Bloomington: Indiana University Press, 1993), 21.

45 Statistics in V. N. Zemskov, "Spetsposelentsy: Po dokumentam NKVD—MVD SSSR," *Sotsiologicheskie issledovaniia* 11 (1990): 3.

46 Statistics in Ivnitskii, *Repressivnaia politika*, 247.

47 Statistics in Ivnitskii, *Golod 1932–1933 godov v SSSR*, 68–69.

CHAPTER 3

1 Quotations from Viktor Kondrashin and Diana [D'Ann] Penner, *Golod: 1931–1933 gody v sovetskoi derevne (na materialakh Povolzh'ia, Dona i Kubani)* (Samara-Penza: Samarskii gos. universitet and Penzenskii gos. Pedagogicheskii universitet im. V. G. Belinskogo, 2002), 117.

2 "An Ogpu Treasurehunt," *Times* (London), March 10, 1932, 12.

3 As quoted in Ivnitskii, *Golod 1932–1933 godov v SSSR*, 258.

4 Statistics in Viola, *The Unknown Gulag*, 103.

5 Quotations from Ivnitskii and Makurov, eds., *Iz istorii raskulachivaniia v Karelii*, 249–50.

6 Quoted in Zelenin, "'Revoliutsiia sverkhu,'" 34.

7 See Ivnitskii, *Golod 1932–1933 godov v SSSR*, 122.

8 Statistics in Davies and Wheatcroft, *The Years of Hunger*, 406.

9 Statistics in Davies and Wheatcroft, *The Years of Hunger*, 435.

10 Statistics in Kondrashin, *Golod 1932–1933 godov*, 108.

11 Quoted in Kondrashin, *Golod 1932–1933 godov*, 97–98.

12 Miron Dolot, *Execution by Hunger: The Hidden Holocaust* (New York: W. W. Norton, 1985), 144.

13 Statistic in Kondrashin, *Golod 1932–1933 godov*, 122.

14 See Ivnitskii, *Golod 1932–1933 godov v SSSR*, 196–97.

15 See L. E. Zelenin, "'Zakon o piati koloskakh': Razrabotka i osushchestvlenie," *Voprosy istorii* 1 (1998): 115.

16 As quoted in Zelenin, "'Zakon o piati koloskakh,'" 118.

17 Statistics in Kondrashin, *Golod 1932–1933 godov*, 142.

18 As quoted in Tracy McDonald, *Face to the Village: The Riazan Countryside under Soviet Rule, 1921–1930* (Toronto: University of Toronto Press, 2011), 182.

19 "Grain Thieves in Russia: Soviet Plans for Guarding Crops," *Times* (London), August 15, 1932, 9.

20 Statistics from Davies and Wheatcroft, *The Years of Hunger*, 102.

21 As quoted in Ivnitskii, *Golod 1932–1933 godov v SSSR*, 143–44.

22 Statistic in Ivnitskii, *Golod 1932–1933 godov v SSSR*, 145.

23 Quotations from Kondrashin and Penner, *Golod*, 197.

24 As quoted in Daniel R. Brower, "Collectivized Agriculture in Smolensk: The Party, the Peasantry, and the Crisis of 1932," *Russian Review* 36, no. 2 (1977): 163.

25 Quoted in Kondrashin, *Golod 1932–1933 godov*, 144.

26 See Ivnitskii, *Repressivnaia politika*, 279.

27 As quoted in Ivnitskii, *Golod 1932–1933 godov v SSSR*, 183.

28 Statistic in Kondrashin, *Golod 1932–1933 godov*, 156.

29 "Food Scarcity in Russia," *Times* (London), December 6, 1932, 13.

30 As quoted in Al'bert Baiburin, "Vvedenie pasportnoi sistemy v SSSR," in *Status dokumenta: Okonchatel'naia bumazhka ili otchuzhdennoe svidetel'stvo?,* ed. I. M. Kaspe (Moscow: NLO, 2012), 79.

31 Statistic and quotation in Zelenin, "Revoliutsiia sverkhu," 33.

32 Statistic and assessment in Zelenin, "Revoliutsiia sverkhu," 40.

33 Statistics in Zelenin, "Revoliutsiia sverkhu," 39–40.

34 Statistics in Iu. A. Moshkov, *Zernovaia problema v gody sploshnoi kollektivizatsii sel'skogo khoziaistva SSSR, 1929–1932* (Moscow: Izd. Moskovskogo universiteta, 1966), 131, 134.

35 As quoted in Sergei Maksudov, "Victory over the Peasantry," in *Hunger by Design: The Great Ukrainian Famine and Its Soviet Context*, ed. Halyna Hryn (Cambridge, MA: Ukrainian Research Institute, 2008), 77–78.

36 Gareth Jones, "Seizure of Land and Slaughter of Stock: Peasants Subsisting on Potatoes and Cattle Fodder," *Western Mail,* April 8, 1933.

37 Neil Prior, "Journalist Gareth Jones' 1935 murder examined by BBC Four," *BBC News,* July 5, 2012, http://www.bbc.com/news/uk-wales-south-east-wales-18691109.

38 Kondrashin, *Golod 1932–1933 godov*, 104.

39 As quoted in Ivnitskii, *Golod 1932–1933 godov v SSSR*, 198.

40 Statistics from Jörg Baberowski, "Stalinismus 'von oben': Kulakendeportationen in der Sowjetunion 1929–1933," *Jahrbücher für Geschichte Osteuropas* 46 (1998): 585.

41 Statistics cited in David Lloyd Hoffmann, *Stalinist Values: The Cultural Norms of Soviet Modernity, 1917–1941* (Ithaca, NY: Cornell University Press, 2003), 73.

42 As quoted in I. E. Zelenin, "Politotdely MTS—prodolzhenie politiki 'chrezvychaishchiny' (1933–1934)," *Otechestvennaia istoriia* 6 (1992): 46.

43 Statistics cited in Davies and Wheatcroft, *The Years of Hunger*, 345.

44 Stalin, *Works*, 13:251.

45 Stalin, *Works*, 13:259.

46 Statistics calculated from Zelenin, "Revoliutsiia sverkhu," 34.

47 "Soviet Farm 'Wreckers,'" *Times* (London), March 13, 1933, 14.

48 "Soviet Plans in Difficulty," *Times* (London), March 28, 1933, 13.

49 Iu. G. Murina, ed., "Sholokhov i Stalin: Perepiska 30-kh godov," *Voprosy istorii* 3 (1994): 14, 15, 16.

50 Stalin to Sholokhov, May 6, 1933, in Murina, ed., "Sholokhov i Stalin," 22.

51 V. V. Kondrashin, "Vykhod iz golodnogo krizisa i ego posledstviia," in *Golod v SSSR, 1929–1934: Dokumenty* (Moscow: Mezhdunarodnyi fond "Demokratiia," 2011–13), 3:41.

52 Statistics in Kondrashin and Penner, *Golod*, 315; S. Uitkroft (Wheatcroft), "O zernovykh balansakh i otsenkakh urozhainosti v SSSR v 1931–1933 gg.," in Danilov et al., eds., *Tragediia sovetskoi derevni*, 3:862.

53 Nicholas Werth, "The Great Ukrainian Famine of 1932–3," trans. David Brown, in *The Routledge History of Genocide*, ed. Cathie Carmichael and Richard C. Maguire (London: Routledge, 2015), 145.

54 As quoted in Kondrashin and Penner, *Golod*, 278.

55 As quoted in Ivnitskii, *Golod 1932–1933 godov v SSSR*, 206.

56 See Kondrashin, *Golod 1932–1933 godov*, 260.

57 See Ivnitskii, *Golod 1932–1933 godov v SSSR*, 205.

58 As quoted in Iu. I. Goncharenko and A. P. Chichenkov, eds., "Golod 1933 goda," *Sovetskie arkhivy* 6 (1990): 49–50.

59 As quoted in Zelenin, "Revoliutsiia sverkhu," 38, 39.

60 Statistics in Ivnitskii, *Golod 1932–1933 godov v SSSR*, 223.

61 Analysis in Ivnitskii, *Repressivnaia politika*, 312.

62 See Kondrashin, *Golod 1932–1933 godov*, 258.

63 Victor Kravchenko, *I Chose Freedom: The Personal and Political Life of a Soviet Official* (New York: C. Scribner's Sons, 1946), 130.

64 V. B. Zhiromskaia, "Golod 1932–1933 gg.: Liudskie poteri," in Kondrashin et al., eds., *Golod v SSSR*, 3:653.

65 See S. Uitkroft (Wheatcroft), "Pokazateli demograficheskogo krizisa v period goloda," in Kondrashin et al., eds., *Golod v SSSR*, 3:767.

66 Kondrashin, *Golod 1932–1933 godov*, 190.

67 See Kondrashin, "Vykhod iz golodnogo krizisa," 47.

68 See Kondrashin, *Golod 1932–1933 godov*, 201.

69 Statistic in Davies and Wheatcroft, *The Years of Hunger*, 263.

70 Statistic in Kondrashin, *Golod 1932–1933 godov*, 200.

71 Statistics in Jasny, *The Socialized Agriculture of the USSR*, 633–34.

72 See Davies and Wheatcroft, *The Years of Hunger*, 326.

73 Statistics in Davies and Wheatcroft, *The Years of Hunger*, 390.

74 Statistic in Kondrashin, *Golod 1932–1933 godov*, 144.

CHAPTER 4

1 Stalin, *Works*, 13:313, 330.
2 Statistics cited in Stephan Merl, *Sozialer Aufstieg im sowjetischen Kolchossystem der 30er Jahre? Über das Schicksal der bäuerlichen Parteimitglieder, Dorfsowjetvorsitzenden, Posteninhaber in Kolchosen, Mechanisatoren und Stachanowleute* (Berlin: Duncker & Humblot, 1990), 92.
3 Statistics supplied in Zelenin, "Politotdely MTS," 49.
4 "Crop Failures in Russia," *Times* (London), May 29, 1934, 13.
5 Lynne Viola, *Peasant Rebels under Stalin: Collectivization and the Culture of Peasant Resistance* (Oxford: Oxford University Press, 1996), 90.
6 As quoted in O. V. Khlevniuk et al., eds., *Stalinskoe Politbiuro v 30-e gody: Sbornik dokumentov* (Moscow: AIRO–XX, 1995), 146.
7 Nicholas Timasheff, *The Great Retreat: The Growth and Decline of Communism in Russia* (New York: E. P. Dutton, 1946).
8 Quotation and statistics from Matthew Lenoe, "Fear, Loathing, Conspiracy: The Kirov Murder as Impetus for Terror," in *The Anatomy of Terror: Political Violence under Stalin*, ed. James Harris (Oxford: Oxford University Press, 2013), 199.
9 See Zelenin, "Politotdely MTS," 57.
10 Statistics supplied by Merl, *Bauern unter Stalin*, 282.
11 Statistic in Merl, *Bauern unter Stalin*, 275.
12 Statistics furnished in Jasny, *The Socialized Agriculture of the USSR*, 395.
13 See Jasny, *The Socialized Agriculture of the USSR*, 400.
14 Statistics in Lewis H. Siegelbaum, *Stakhanovism and the Politics of Productivity in the USSR, 1935–1941* (Cambridge: Cambridge University Press, 1988), 63.
15 J. V. Stalin, *Works* (London: Red Star Press, 1978), 14:98.
16 Stalin, *Works*, 14:382.
17 Statistics cited in Walter Scott Dunn, *The Soviet Economy and the Red Army, 1930–1945* (Westport, CT: Praeger, 1995), 230.
18 Statistics found in Alec Nove, *An Economic History of the U.S.S.R.*, 2d ed. (London: Penguin Books, 1989), 250.
19 Statistics supplied in Kondrashin and Penner, *Golod*, 355.
20 Statistic from D. Gale Johnson, "Agriculture," in *The Soviet Union Today: An Interpretive Guide*, 2d ed., ed. James Cracraft (Chicago: University of Chicago Press, 1988), 199.
21 Quoted from Merl, *Bauern unter Stalin*, 255.
22 As quoted in Kondrashin, *Golod 1932–1933 godov*, 304.
23 Statistics in Kondrashin, *Golod 1932–1933 godov*, 306.
24 Statistics in Merl, *Bauern unter Stalin*, 420.
25 See Sue Bridger, "The Heirs of Pasha: The Rise and Fall of the Soviet Woman Tractor Driver," in *Gender in Russian History and Culture*, ed. Linda Edmondson (New York: Palgrave Macmillan, 2001), 202.
26 Davies, "Making Economic Policy," 68.
27 Quoted in Merl, *Bauern unter Stalin*, 431.
28 Statistics in Merl, *Bauern unter Stalin*, 208.
29 See Merl, *Bauern unter Stalin*, 251.
30 Zemskov, "'Kulatskaia ssylka' v 30-e gody," 10:14.
31 See Ivnitskii, *Sud'ba raskulachennykh v SSSR*, 281.
32 David R. Shearer and Vladimir Khaustov, *Stalin and the Lubianka: A Documentary History of the Political Police and Security Organs in the Soviet Union, 1922–1953* (New Haven, CT: Yale University Press, 2009), 195.
33 This extraordinary document has been reproduced in Shearer and Khaustov, *Stalin and the Lubianka*, 200–205.
34 Quotation and statistics in Sheila Fitzpatrick, *Stalin's Peasants: Resistance and Survival in the Russian Village after Collectivization* (New York: Oxford University Press, 1994), 202.
35 For this terrible statistic, see J. Arch Getty, Gábor T. Rittersporn, and Viktor N. Zemskov, "Victims of the Soviet Penal System in the Pre-War Years: A First Approach on the Basis of Archival Evidence," *American Historical Review* 98, no. 4 (October 1993): 1022.
36 See Ivnitskii, *Sud'ba raskulachennykh v SSSR*, 146.
37 Statistic in Ivnitskii, *Sud'ba raskulachennykh v SSSR*, 77.
38 Statistic cited from Zemskov, "'Kulatskaia ssylka' v 30-e gody," 14.
39 Statistics from Zemskov, "'Kulatskaia ssylka' v 30-e gody," 7.
40 See Viola, *The Unknown Gulag*, 179.

CONCLUSION

1 Peter Holquist, "Conduct Merciless Mass Terror: Decossackization on the Don, 1919," *Cahiers du monde russe* 38, nos. 1–2 (January–June 1997): 134.

2 As quoted in Goldman, "Hunger and Politics," 40.

3 Jörg Baberowski, "Die Kollektivierung der Landwirtschaft und der Terror gegen die Kulaken," in *Europa und die Europäer: Quellen und Essays zur modernen europäischen Geschichte*, ed. Rüdiger Hohls, Iris Schröder, and Hannes Siegrist (Stuttgart: Franz Steiner Verlag, 2005), 318. In a more recent book, Baberowski emphasizes Stalin's pathologies over his modernizing project. See *Scorched Earth: Stalin's Reign of Terror*, trans. Steven Gilbert, Ivo Komljen, and Samantha Jeanne Taber (New Haven, CT: Yale University Press, 2016).

4 Viola, *Peasant Rebels under Stalin*, 92.

5 Diane P. Koenker and Ronald D. Bachman, eds., *Revelations from the Russian Archives: Documents in English Translation* (Washington, DC: Library of Congress, 1997), 412.

6 Stalin, *Works*, 13:194.

7 Andrzej Walicki, *Marxism and the Leap to the Kingdom of Freedom: The Rise and Fall of the Communist Utopia* (Stanford, CA: Stanford University Press, 1995).

8 Frank Dikötter, *Mao's Great Famine: The History of China's Most Devastating Catastrophe, 1958–62* (London: Bloomsbury, 2010), xii.

9 Paul R. Bartrop and Steven Leonard Jacobs, eds., *Modern Genocide: The Definitive Resource and Document Collection* (Santa Barbara, CA: ABC-CLIO, 2015), 4:2062.

10 Council of Europe, Parliamentary Assembly, *Adopted Texts, 2010, Ordinary Session* (Second Part) (Strasbourg: Council of Europe Publishing, 2010), 49.

11 Ella Maksimova, "Istorik Viktor Kondrashin: "Ne Rossiia ubivala Ukrainu. Vozhd'—svoi narod," *Izvestiia*, October 22, 2008, https://web.archive.org/web/20101201204942; http://izvestia.ru/hystory/article3121813.

12 Statistics supplied in Kondrashin, *Golod 1932–1933 godov*, 246.

13 See the statistical table in Viola et al., eds., *The War against the Peasantry*, 364.

14 Kondrashin, *Golod 1932–1933 godov*, 247–48.

15 Quotation from A. Latyshev, "Stalin i kino," in *Surovaia drama naroda: Uchenye i publitsisty o prirode slalinizma*, ed. Iu. P. Senokosov (Moscow: Politizdat, 1989), 503.

16 *Sel'skoe khoziaistvo SSSR: Ezhegodnik 1935* (Moscow: Sel'khozgiz, 1936), 256.

17 Holland Hunter and Janusz M. Szyrmer, *Faulty Foundations: Soviet Economic Policies, 1928–1940* (Princeton, NJ: Princeton University Press, 1992), 241–54.

18 A. A. Barsov, "NEP i vyravnivanie ekonomicheskikh otnoshenii mezhdu gorodom i derevnei," in *Novaia ekonomicheskaia politika: Voprosy teorii i istorii*, ed. M. P. Kim et al. (Moscow: Nauka, 1974), 103.

19 *Sotsialisticheskoe stroitelstvo SSSR: Statisticheskii ezhegodnik* (Moscow: TsUNKhU Gosplana SSSR, 1935), 302, 324–25.

20 Statistics in Jasny, *The Socialized Agriculture of the USSR*, 441.

21 See US Bureau of the Census, *USA/USSR: Facts and Figures* (Washington, DC: US Government Printing Office, 1991), tables 3.1, 3.2, 6.1–6.17. The Soviet figure includes an undisclosed proportion of collective farmers "occupied in non-agricultural branches of collective farms (such as industry, fish-catching, construction, trade, and services), but excludes family members working in private subsidiary agriculture."

22 On this question, see V. V. Kondrashin, "Golod v krest'ianskom mentalitete," in *Mentalitet i agrarnoe razvitie Rossii, XIX–XX vv.*, ed. V. P. Danilov and L. V. Milov (Moscow: ROSSPEN, 1996), 115–23.

23 The memorandum has been reproduced in Oleg V. Khlevniuk, *The History of the GULAG: From Collectivization to the Great Terror*, trans. Vadim A. Staklo, ed. David J. Nordlander (New Haven, CT: Yale University Press, 2004), 288–89.

24 Statistics in Neil J. Melvin, *Soviet Power and the Countryside: Policy Innovation and Institutional Decay* (New York: Palgrave Macmillan, 2003), 5.

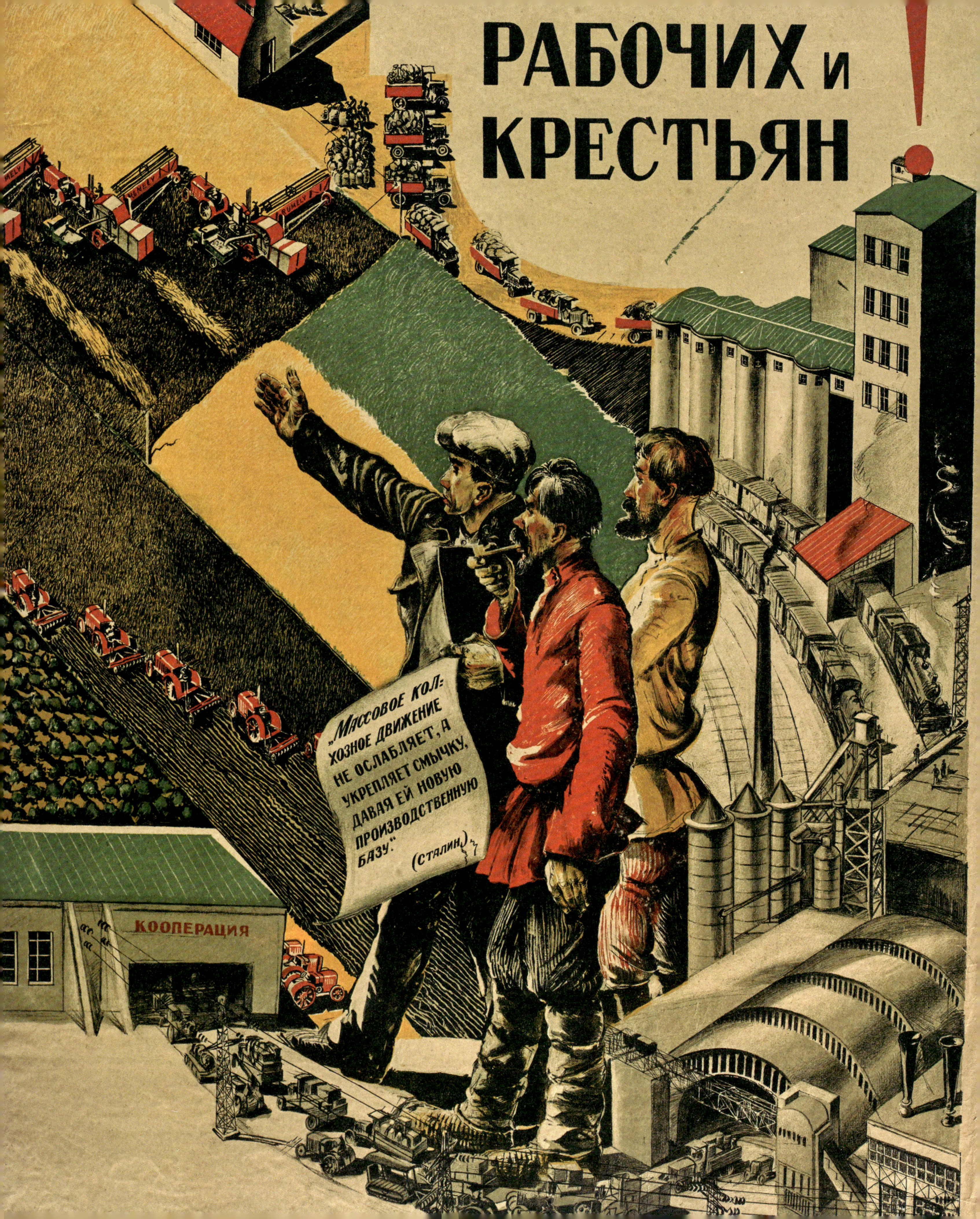
РАБОЧИХ и
КРЕСТЬЯН
„Массовое кол-
хозное движение
не ослабляет, а
укрепляет смычку,
давая ей новую
производственную
базу."
(СТАЛИН)
КООПЕРАЦИЯ

GLOSSARY

Article 58 of the criminal code Counterrevolution and anti-Soviet activity

Article 61 of the criminal code Failure to fulfill obligations to the state

Article 107 of the criminal code Speculation, withholding grain

bab'i bunty Women's rebellions during collectivization

centner (quintal) 100 kilograms or one-tenth of a metric ton

Central Committee The most senior Communist Party officials

Cheka The Extraordinary Commission, or secret police (1917–22)

class In the Marxist conception, an immutable social category

dekulakization A murderous campaign against peasants resisting collectivization

fivefold tax A punishment imposed for tax arrears starting in summer 1929

Five-Year Plan The central economic planning framework adopted in 1928

FSB The Federal Security Service of the Russian Federation

Glavlit The censorship authority

Gosplan The central economic planning agency

GPU The Main Political Administration, or secret police (1922–23)

gulag The network of forced-labor camps (from the acronym for *Glavnoe upravlenie ispravitel'no-trudovykh lagerei*, or Main Administration for Corrective Labor Camps)

The Gulag Archipelago A book by Alexander Solzhenitsyn about the Soviet forced-labor camp system

KGB The Committee of State Security of the Soviet Union

Komsomol The Communist Youth League

"kulak" A well-to-do peasant or any rural peasant

labor day A complicated means for calculating the wages of collective farmers

Machine Tractor Stations (MTS) Offices that lent out agricultural machinery

middle peasants In the Marxist conception, the bulk of the peasants, neither rich nor poor

◀ Detail of "Raise the banner of the productive *smychka* of workers and peasants!," *see page 122.*

NEPmen Private entrepreneurs during the period of the NEP

New Economic Policy (NEP) A program temporarily relaxing antimarket policies in 1921–28

NGOs Non-governmental organizations

NKVD The People's Commissariat of Internal Affairs, or secret police (1934–46)

OGPU The Unified Main Political Administration, or secret police (1923–34)

Party Congress A meeting of Communist Party officials convened to make major decisions

peasant A traditional farmer

People's Commissariat Central government department

***peregiby* (lit., excesses)** Violence and abuse

plenum A thrice-yearly meeting of Central Committee members

Podkulachniki "Kulak henchmen," allegedly doing the bidding of "kulaks"

Politburo The top Communist Party leaders

political sections (*politotdely*) Agencies set up in 1932 to watch over local rural officials

pood An old Russian unit of measure equal to 36 pounds

Pravda The flagship newspaper of the Russian Communist Party

Presidium of the Central Committee Executive board of the Central Committee

proletariat The class of industrial workers

Reds The Communists

smychka The union of workers and peasants

socialization State control of property

Sovnarkom The Council of People's Commissars, or the Soviet government

speculation The act of engaging in normal market relations

state farms Large farms run like agricultural factories

troika A committee of three officials, typically representing the party, state, and secret police

tsentner *See* centner

udarnik/i Shock worker(s), the most dedicated worker(s)

Ural-Siberian method A method relying on social pressure to extract grain from peasants

Whites Anti-Bolshevik forces during the Russian Civil War

ZAGs The state civil registry office

▶ Detail of "Join the ranks of the shock worker brigades," *see page 63*.

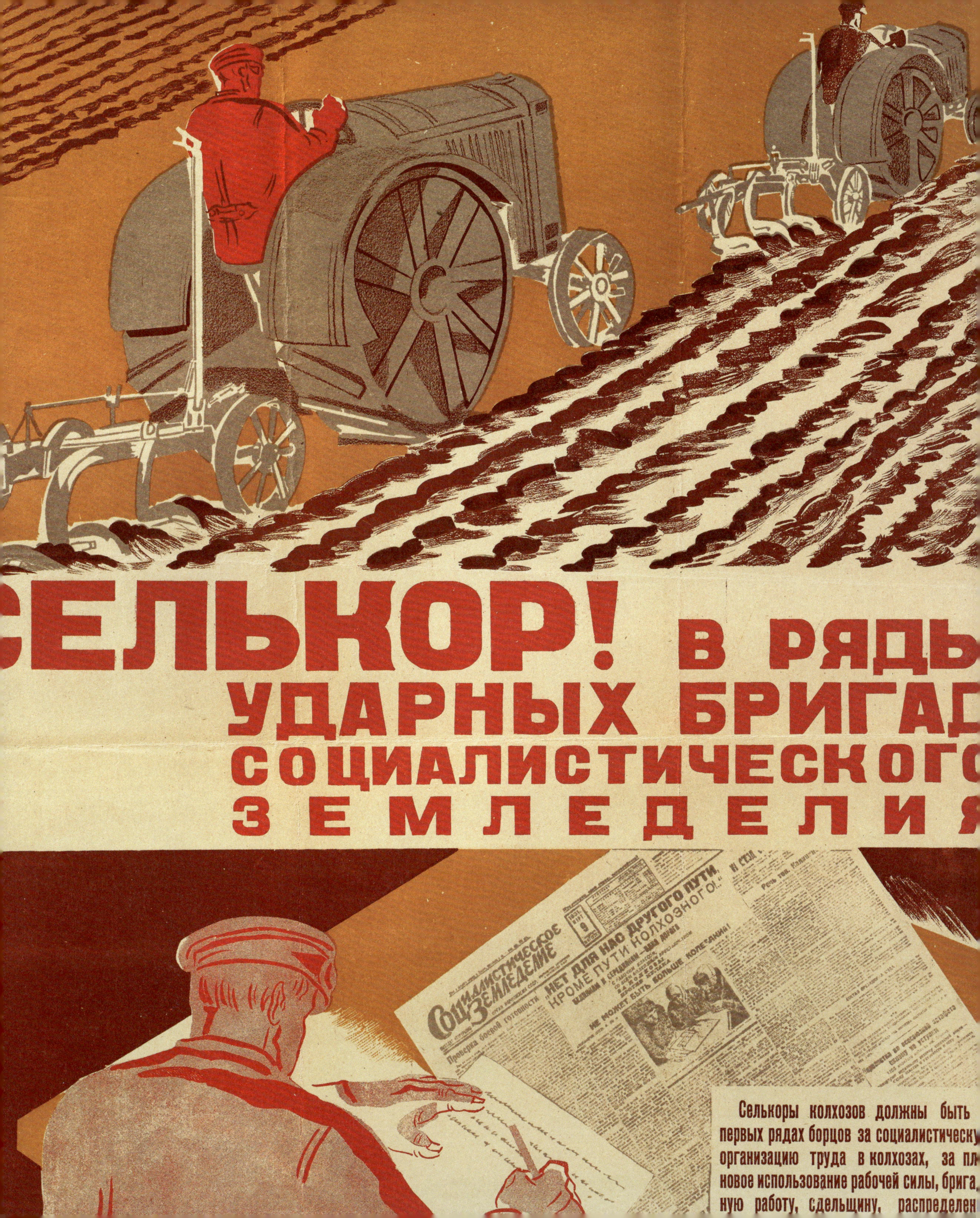
СЕЛЬКОР! В РЯД
УДАРНЫХ БРИГАД
СОЦИАЛИСТИЧЕСКОГО
ЗЕМЛЕДЕЛИЯ
Социалистическое Земледелие
НЕТ ДЛЯ НАС ДРУГОГО ПУТИ, КРОМЕ ПУТИ КОЛХОЗНОГО!
Селькоры колхозов должны быть
первых рядах борцов за социалистическу
организацию труда в колхозах, за пл
новое использование рабочей силы, брига
ную работу, сдельщину, распределен

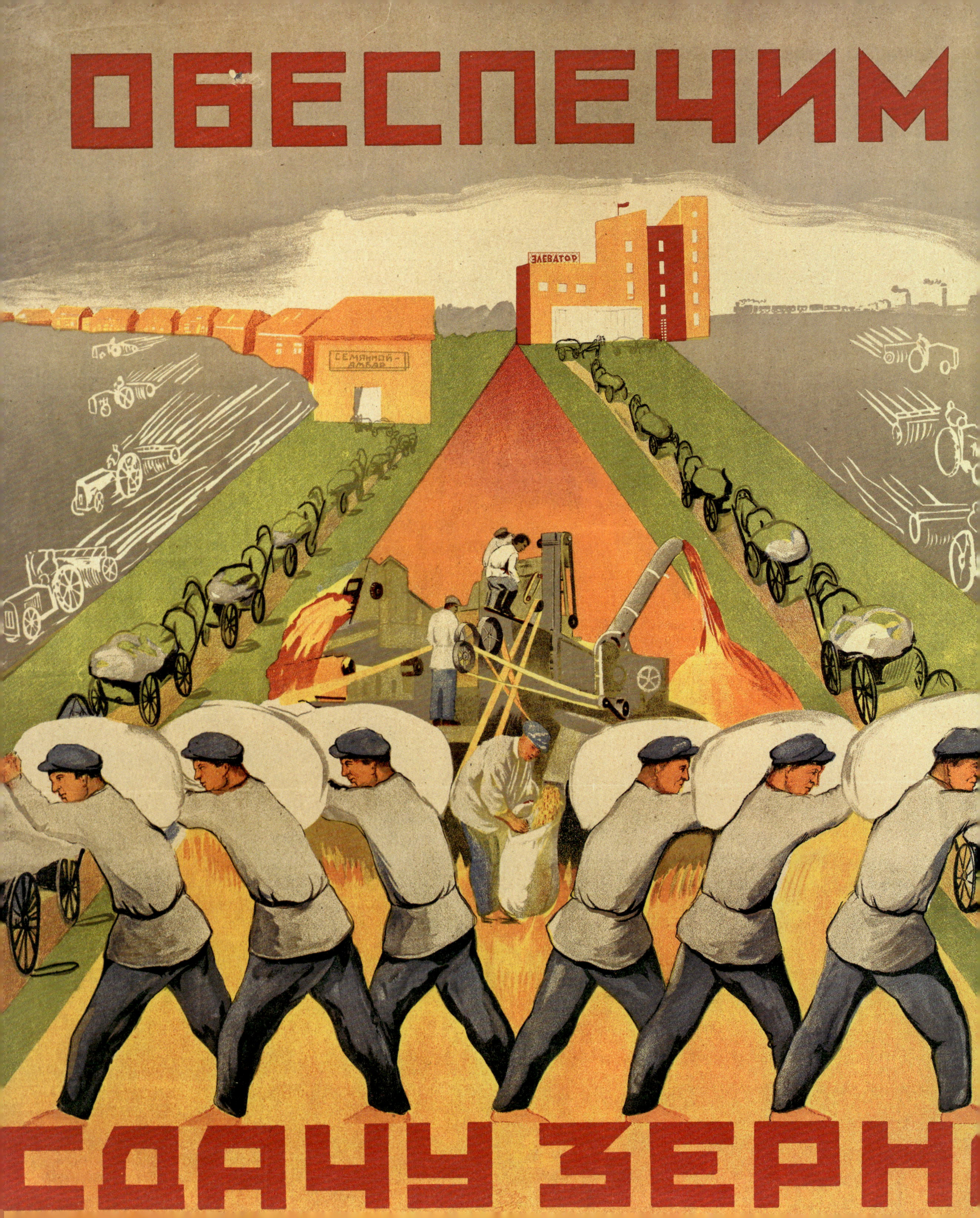
ОБЕСПЕЧИМ
ЭЛЕВАТОР
СЕМЯННОЙ-АМБАР
СДАЧУ ЗЕРН

FURTHER READING

Readers interested in learning more about Soviet collectivization and related topics discussed in this book will find the most important English-language scholarship listed below.

PRIMARY SOURCES

Dolot, Miron. *Execution by Hunger: The Hidden Holocaust*. New York: W. W. Norton, 1985.

Grigorenko, Petro G. *Memoirs*. Translated by Thomas P. Whitney. New York: W. W. Norton, 1982.

Kravchenko, Victor. *I Chose Freedom: The Personal and Political Life of a Soviet Official*. New York: C. Scribner's Sons, 1946.

Leshuk, Leonard, ed. *Days of Famine, Nights of Terror: Firsthand Accounts of Soviet Collectivization, 1928–1934*. 2d ed. Translated by Raimund Rueger. Washington, DC: Europa University Press, 2000.

Litinenko, Olga, and James Riordan, eds. *Memories of the Dispossessed: Descendants of Kulak Families Tell Their Stories*. Nottingham, UK: Bramcote Press, 1998.

Storella, C. J., ed. *The Voice of the People: Letters from the Soviet Village, 1918–1932*. New Haven, CT: Yale University Press, 2013.

Viola, Lynne, et al., eds. *The War against the Peasantry, 1927–1930: The Tragedy of the Soviet Countryside*. New Haven, CT: Yale University Press, 2005.

SECONDARY SOURCES

Allen, Robert C. *Farm to Factory: A Reinterpretation of the Soviet Industrial Revolution*. Princeton, NJ: Princeton University Press, 2003.

Baberowski, Jörg. *Scorched Earth: Stalin's Reign of Terror*. Translated by Steven Gilbert, Ivo Komljen, and Samantha Jeanne Taber. New Haven, CT: Yale University Press, 2016.

Belov, Fedor. *The History of a Soviet Collective Farm*. New York: Frederick A. Prager, 1955.

Buckley, Mary. *Mobilizing Soviet Peasants: Heroines and Heroes of Stalin's Fields*. Lanham, MD: Rowman and Littlefield, 2006.

Conquest, R. *Harvest of Sorrow: Soviet Collectivization and the Terror-Famine*. New York: Oxford University Press, 1986.

Davies, R. W. *The Industrialization of Soviet Russia*, vol. 1, *The Socialist Offensive: The Collectivization of Soviet Agriculture, 1929–1930*, and vol. 2, *The Soviet Collective Farm, 1929–1930*. Cambridge, MA: Harvard University Press, 1980.

———. "Making Economic Policy." In *Behind the Façade of Stalin's Command Economy: Evidence from the Soviet State*

◀ Detail of "We will ensure the delivery of grain," *see page 30*.

and Party Archives, edited by Paul R. Gregory, 61–80. Stanford, CA: Hoover Institution Press, 2001.

Davies, R. W., and Stephen G. Wheatcroft. *The Years of Hunger: Soviet Agriculture, 1931–1933*. New York: Palgrave Macmillan, 2004.

Fitzpatrick, Sheila. "The Great Departure: Rural-Urban Migration in the Soviet Union, 1929–1933." In *Social Dimensions of Soviet Industrialization*, edited by William R. Rosenberg and Lewis H. Siegelbaum, 15–40. Bloomington: Indiana University Press, 1993.

———. *Stalin's Peasants: Resistance and Survival in the Russian Village after Collectivization*. New York: Oxford University Press, 1994.

Getty, J. Arch. *Practicing Stalinism: Bolsheviks, Boyars, and the Persistence of Tradition*. New Haven, CT: Yale University Press, 2013.

Goldman, Wendy. "Hunger and Politics: The Prelude to the Great Terror." In *Hunger and Scarcity under State-Socialism*, edited by M. Middell and F. Wemheuer, 31–58. Leipzig: Leipziger Universitatsverlag, 2012.

Graziosi, Andrea. *The Great Soviet Peasant War: Bolsheviks and Peasants, 1918–1934*. Cambridge, MA: Ukrainian Research Institute, 1996.

Hagenloh, Paul. *Stalin's Police: Public Order and Mass Repression in the USSR, 1926–1941*. Baltimore: The John Hopkins University Press, 2009.

Harris, James R. "The Growth of the Gulag: Forced Labor in the Urals Region, 1929–1931." *Russian Review* 56 (April 1997): 265–80.

Heinzen, James W. *Inventing a Soviet Countryside: State Power and the Transformation of Rural Russia, 1917–1929*. Pittsburgh: Pittsburgh University Press, 2004.

Hernandez, Richard L. "Sacred Sound and Sacred Substance: Church Bells and the Auditory Culture of the Russian Village during the Bolshevik 'Velikii Perelom.'" *American Historical Review* 109 (2004): 1475–1504.

———. "Good Shepherds: The Public Authority of Parish Clergy in the Era of Collectivization." *Russian History* 32 (2005): 195–214.

Hudson, Hugh D., Jr. *Peasants, Political Police, and the Early Soviet State: Surveillance and Accommodation under the New Economic Policy*. New York: Palgrave Macmillan, 2012.

Hughes, James. *Stalin, Siberia, and the Crisis of the New Economic Policy*. Cambridge: Cambridge University Press, 1991.

———. "Capturing the Russian Peasantry: Stalinist Grain Procurement Policy and the 'Ural-Siberian Method.'" *Slavic Review* 53, no. 1 (1994): 76–103.

———. *Stalinism in a Russian Province: A Study of Collectivization and Dekulakization in Siberia*. New York: St. Martin's Press, 1996.

———. "Re-evaluating Stalin's Peasant Policy in 1928–30." In *Transforming Peasants: Society, State and the Peasantry, 1861–1930*, edited by Judith Pallot. New York: St. Martin's Press, 1998.

Hunter, Holland, and Janusz M. Szyrmer. *Faulty Foundations: Soviet Economic Policies, 1928–1940*. Princeton, NJ: Princeton University Press, 1992.

Jasny, Naum. *The Socialized Agriculture of the USSR: Plans and Performance*. Stanford, CA: Stanford University Press, 1949.

Khlevnyuk, Oleg. "The Economy of the Gulag." In *Behind the Façade of Stalin's Command Economy: Evidence from the Soviet State and Party Archives*, edited by Paul R. Gregory, 111–30. Stanford, CA: Hoover Institution Press, 2001.

Kindler, Robert. "On the Run—the Kazakh Nomads and the Famine, 1930–1934." In *Hunger and Scarcity under State-Socialism*, edited by M. Middell and F. Wemheuer, 59–82. Leipzig: Leipziger Universitatsverlag, 2012.

Kuromiya, Hiroaki. "The Soviet Famine of 1932–1933 Reconsidered." *Europe-Asia Studies* 60, no. 4 (June 2008): 663–75.

Laird, Roy D. "Kolkhozy, the Russian Achilles Heel: Failed Agrarian Reform." *Europe-Asia Studies* 49, no. 3 (May 1997): 469–78.

Lewin, Moshe. *Russian Peasants and Soviet Power: A Study of Collectivization*. Translated by Irene Nove with John Biggart. Preface by Alec Nove. Evanston, IL: Northwestern University Press, 1968.

Maksudov, Sergei. "Victory over the Peasantry." In *Hunger by Design: The Great Ukrainian Famine and Its Soviet Context*, edited by Halyna Hryn, 53–101. Cambridge, MA: Ukrainian Research Institute, 2008.

Manning, Roberta T. "The Rise and Fall of the Extraordinary Measures, January–June, 1928: Towards a Reassessment of the Onset of the Stalin Revolution." *Carl Beck Papers in Russian and East European Studies* 1504 (January 2001).

McDonald, Tracy. *Face to the Village: The Riazan Countryside under Soviet Rule, 1921–1930*. Toronto: University of Toronto Press, 2011.

Naimark, Norman M. *Stalin's Genocides*. Princeton, NJ: Princeton University Press, 2010.

Penner, D'Ann R. "Stalin and the Ital'ianka of 1932–1933 in the Don Region." *Cahiers du monde russe* 39, nos. 1–2 (January–June 1998): 27–67.

Reese, Roger R. "Red Army Opposition to Forced Collectivization, 1929–1930: The Army Wavers." *Slavic Review* 55 (Spring 1996): 24–45.

Scott, James C. *Weapons of the Weak: Everyday Forms of Peasant Resistance*. New Haven, CT: Yale University Press, 1985.

———. *Seeing Like a State: How Certain Schemes to Improve the Human Condition Have Failed*. New Haven, CT: Yale University Press, 1998.

Shearer, David R. *Policing Stalin's Socialism: Repression and Social Order in the Soviet Union, 1924–1953*. New Haven, CT: Yale University Press, 2009.

Solomon, Peter H. *Soviet Criminal Justice under Stalin*. Cambridge: Cambridge University Press, 1996.

Viola, Lynne. *Peasant Rebels under Stalin: Collectivization and the Culture of Peasant Resistance*. New York: Oxford University Press, 1996.

———. *The Unknown Gulag: The Lost World of Stalin's Special Settlements*. Oxford: Oxford University Press, 2007.

———. "The Aesthetic of Stalinist Planning and the World of the Special Villages." In *Peopling the Russian Periphery: Borderland Colonization in Eurasian History*, edited by Nicholas B. Breyfogle, Abby Schrader, and Willard Sunderland, 189–212. New York: Routledge, 2007.

———. "Stalin's Empire: The Gulag and Police Colonization in the Soviet Union in the 1930s." In *Stalin and Europe: Imitation and Domination, 1928–1953*, edited by Timothy Snyder and Ray Brandon, 18–43. New York: Oxford University Press, 2014.

▶ *following:* Detail of "We pledge in 1935 to produce a great harvest," *see page 76*.

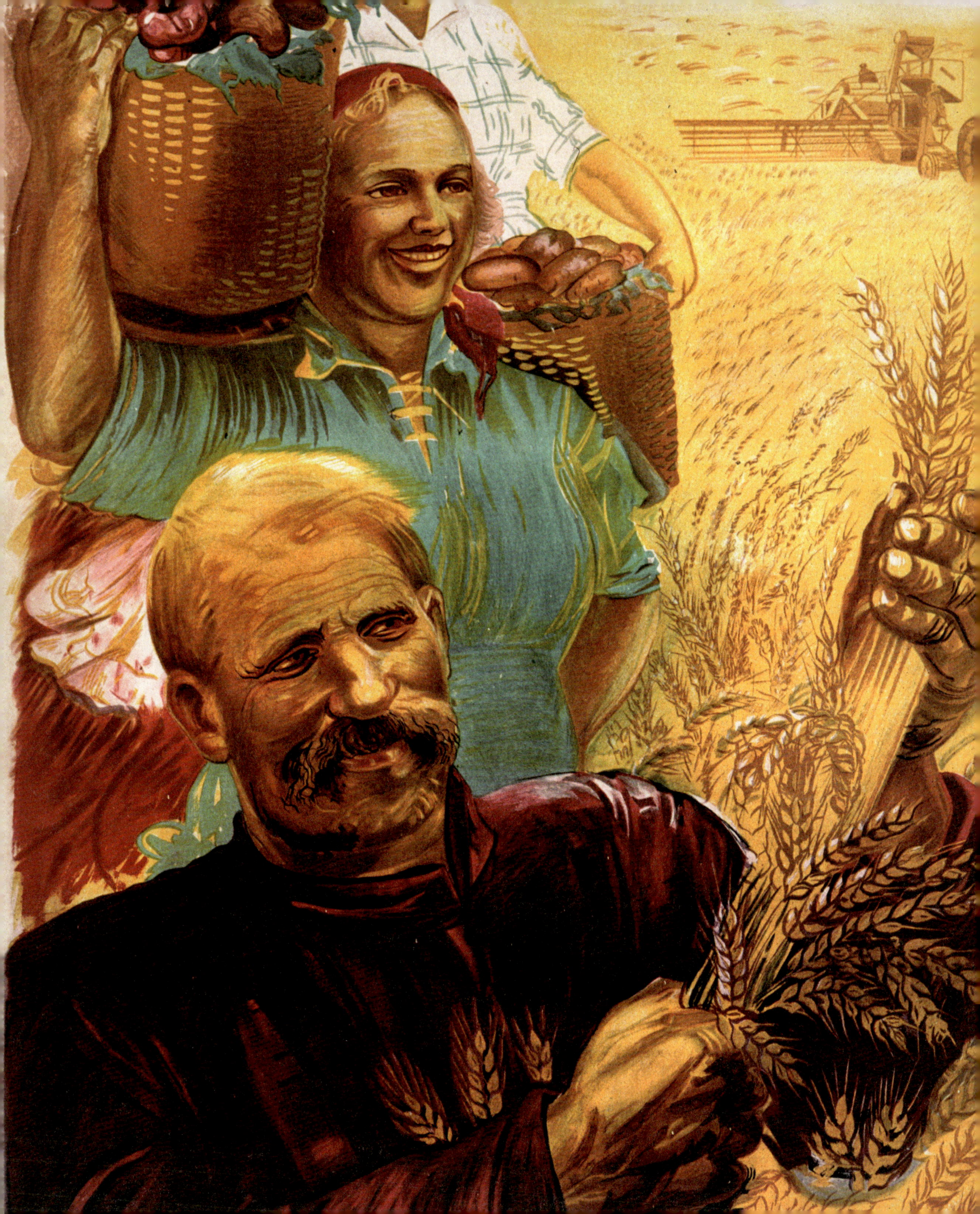

INDEX

Entries followed by an "m" indicate maps.

▶ Detail of "The land is the common and equal property," *see page 5*.

СТРО

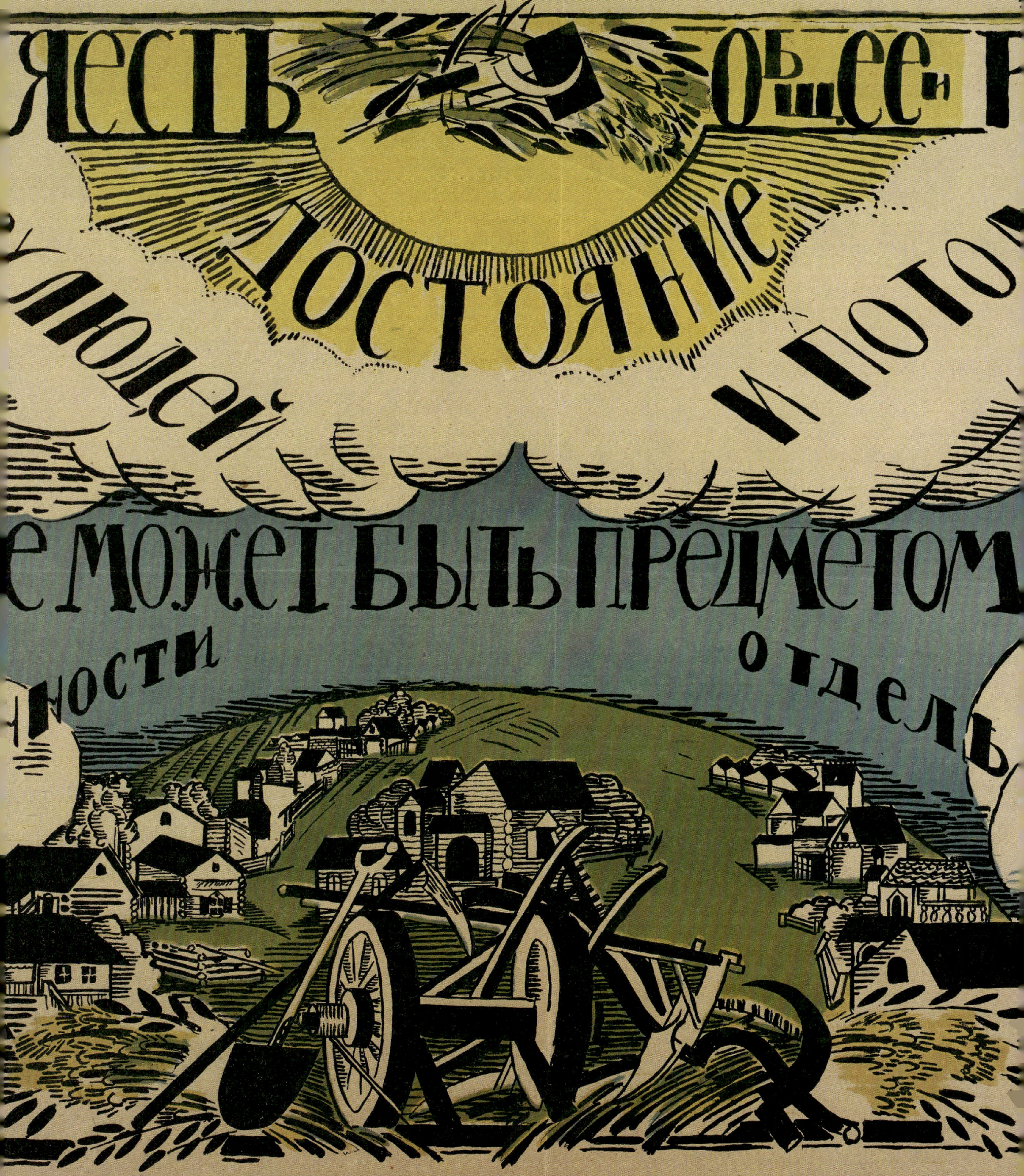

Я ЕСТЬ
ОБЩЕЕ И
ДОСТОЯНИЕ
ЛЮДЕЙ
И ПОТОМ
Е МОЖЕТ БЫТЬ ПРЕДМЕТОМ
НОСТИ
ОТДЕЛЬ